THE SUN IS RISING
IN AFRICA AND THE MIDDLE EAST

Pan Stanford Series on Renewable Energy

Series Editor
Wolfgang Palz

Vol. 1
Power for the World: The Emergence of Electricity from the Sun
Wolfgang Palz, ed.
2010
978-981-4303-37-8 (Hardcover)
978-981-4303-38-5 (eBook)

Vol. 2
Wind Power for the World: The Rise of Modern Wind Energy
Preben Maegaard, Anna Krenz, and Wolfgang Palz, eds.
2013
978-981-4364-93-5 (Hardcover)
978-981-4364-94-2 (eBook)

Vol. 3
Wind Power for the World: International Reviews and Developments
Preben Maegaard, Anna Krenz, and Wolfgang Palz, eds.
2013
978-981-4411-89-9 (Hardcover)
978-981-4411-90-5 (eBook)

Vol. 4
Solar Power for the World: What You Wanted to Know about Photovoltaics
Wolfgang Palz, ed.
2013
978-981-4411-87-5 (Hardcover)
978-981-4411-88-2 (eBook)

Vol. 5
Sun above the Horizon: Meteoric Rise of the Solar Industry
Peter F. Varadi
2014
978-981-4463-80-5 (Hardcover)
978-981-4613-29-3 (Paperback)
978-981-4463-81-2 (eBook)

Vol. 6
Biomass Power for the World: Transformations to Effective Use
Wim van Swaaij, Sascha Kersten, and Wolfgang Palz, eds.
2015
978-981-4613-88-0 (Hardcover)
978-981-4669-24-5 (Paperback)
978-981-4613-89-7 (eBook)

Vol. 7
The U.S. Government & Renewable Energy: A Winding Road
Allan R. Hoffman
2016
978-981-4745-84-0 (Paperback)
978-981-4745-85-7 (eBook)

Vol. 8
Sun towards High Noon: Solar Power Transforming Our Energy Future
Peter F. Varadi
2017
978-981-4774-17-8 (Paperback)
978-1-315-19657-2 (eBook)

Vol. 9
The Sun Is Rising in Africa and the Middle East: On the Road to a Solar Energy Future
Peter F. Varadi, Frank Wouters, and Allan R. Hoffman
2018
978-981-4774-89-5 (Paperback)
978-1-351-00732-0 (eBook)

Vol. 10
The Triumph of the Sun: The Energy of the New Century
Wolfgang Palz
2018

Pan Stanford Series on Renewable Energy – Volume 9

THE SUN IS RISING
IN AFRICA AND THE MIDDLE EAST

On the Road to a Solar Energy Future

Peter F. Varadi | Frank Wouters | Allan R. Hoffman

Series Editor
Wolfgang Palz

Contributors
Anil Cabraal
Richenda Van Leeuwen

PAN STANFORD PUBLISHING

Published by

Pan Stanford Publishing Pte. Ltd.
Penthouse Level, Suntec Tower 3
8 Temasek Boulevard
Singapore 038988

Email: editorial@panstanford.com
Web: www.panstanford.com

British Library Cataloguing-in-Publication Data
A catalogue record for this book is available from the British Library.

The Sun Is Rising in Africa and the Middle East: On the Road to a Solar Energy Future

ISBN 978-981-4774-89-5 (Paperback)
ISBN 978-1-351-00732-0 (eBook)

Contents

Preface

It is hard not to be excited about the spectacular growth of the solar industry in recent years. Unthinkable a few years ago, millions of private people have solar systems on their homes, businesses have solar rooftops, and utilities are now routinely building solar projects that are larger than 1 GW in size. The extent of global exponential growth keeps surprising us year upon year and even month upon month. This book is the latest in a series of books on solar energy by the authors. Peter F. Varadi's book *Sun above the Horizon*, by a true solar energy pioneer, takes you on a trip with him through the early days of solar, and his following book *Sun towards High Noon* describes the recent meteoric rise of solar installations. While working on the latter book, it dawned upon us that the great promise of solar energy was finally materializing in Africa and the Middle East. But we also realized that nobody had really investigated these rapidly emerging markets in a systematic way, so we decided to write this book.

The challenge of writing a book about such a dynamic sector quickly became evident. While we were writing it, several world records on solar energy deployment were broken in the Middle East, requiring us to update our chapters continuously. We are sure that shortly after the book is printed, new world records will be set, and more ambitious solar targets will be announced.

But it is not only large projects that are noteworthy. We are equally excited about the ability of solar to improve human welfare by providing high-quality lighting, TV, Internet access and phone charging for Africa's rural population, access to clean water, improved educational opportunities, and enhanced access to quality medical care. Affordable PV panels, Li-ion batteries, efficient LED lights, and mobile money have combined into a sustainable business model that has already served hundreds of thousands of families in Kenya, Uganda, and Tanzania and is spreading rapidly across Africa. Supermarkets in Abu Dhabi now stock affordable solar equipment, and migrant workers from Africa and Asia buy these products and send them home.

The research we did for this book involved many people in our network, and we are very grateful for their knowledge, wisdom, input, and support. Special thanks are due for Richenda Van Leeuwen, Anil Cabraal, and Wolfgang Palz for their contributions. Their lifelong dedication to energy for development truly reflects in their chapters, which are an indispensable part of this book.

We hope you are stimulated to support and further advance the solar (and renewable) energy revolution currently unfolding before our eyes.

Introduction

The share of solar electricity in Africa and the Middle East in 2017 is still very small, but it is expected that these areas will become the next large markets for solar electric systems. This book provides a review of why and how this will happen.

Solar electricity has become a serious part of *global* electricity production in a short time. In 2010, the *globally* installed solar electric (PV) capacity was only 10 gigawatt (GW). Seven years later, by the end of 2016, the *globally* installed solar electric generating capacity was 296 GW.[1] After 7 years of astronomic growth, the installed solar electric capacity now produces about as much electricity as half of all of the U.S. nuclear electric power plants combined.

On the other hand, the installed solar electric capacity in Africa in 2016 was only 2.9 GW, 0.98% of the globally installed PV capacity and in the Middle East only 1.5 GW, 0.5% of the globally installed PV capacity. How small these amounts are can be seen if we compare them with the installed capacity in Australia. Africa has about 53 times as many inhabitants as Australia, but in Australia 6.3 GW solar electricity is in operation, which is two times as much as in Africa.

In Africa and the Middle East, the utilization of solar energy to generate electricity for various reasons has begun to rise and will continue to grow tremendously. Obviously, one of the reasons is that the cost of solar electricity has dropped incredibly and made the produced electricity less expensive than that produced by traditional utilities, especially using diesel or nuclear. Other reasons are the growing demand for electricity, flexibility to deploy a solar electricity–generating system anywhere, dependability, and no need for the expensive transportation of fuel.

Africa has been called the "lost continent," but since 2000, it is not lost anymore. Many of its countries are among the world's fastest-growing economies and are on the way to becoming middle-

---

[1]International Renewable Energy Agency (IRENA).

income countries, with a growing middle class. The extreme poverty is declining and the entire continent is trying to leapfrog from the Iron Age to the Electric Age. It is an astonishing fact that Africa's "poor population" having still no access to electricity is spending vast amounts of money on kerosene for lights and taking their mobile phones to stores for recharging. They could use that money to switch to solar electricity. This is now being understood by more and more people, and many are switching to solar. Many African countries and donors from America, Europe, and Asia are realizing that installing utility-scale solar systems would provide the needed electricity faster and more securely than building many small-size hydroelectric systems or polluting with diesel generator sets requiring very expensive fuel.

The countries in the Middle East, many of which possess an incredible amount of oil and gas, have started deploying large solar electric systems to produce their electricity. Those countries, including Saudi Arabia, had not realized until now that they needed them. However, there is growing recognition that without utilizing solar energy to produce electricity, in a few years they will face serious economic problems or even bankruptcy.

1

Solar Energy in Africa and in the Middle East

1.1 An Overview of Energy Production and Consumption in Africa and the Middle East

1.1.1 Africa

Africa is the world's second largest continent, its land area being second only to Asia. With a population of just over 1.2 billion, about one-sixth of the world's total, it is second only to Asia as well, which has a population of 4.4 billion. Surrounded by water on all sides, it has 54 sovereign states and 10 non-sovereign territories. Parts of France, Italy, Portugal, and Spain are located on the continent. Egypt, although extending into Asia through the Sinai Peninsula, is considered an African state.

The African population is the youngest among all the continents—more than half of Africans are less than 20 years old—and by 2050, the number of Africans is expected to exceed 2 billion. It is a continent with serious problems as well as significant potential for addressing many of these problems in the decades ahead. Critical to this potential is the development of Africa's energy resources.

Energy production and consumption varies widely across the continent, with some African countries exporting energy to other African countries or the global market, while others lack even basic energy infrastructure. Overall, the African continent is a net energy exporter, exporting about 40% of its energy production. It accounts for about one-eighth of the world's oil production and 7% of natural gas production. The World Bank estimates that the African economy is about the size of the Netherlands' economy—approximately one-sixth that of the U.S.—and that Africa's infrastructure is "...the most deficient and costly in the developing world." Energy development has not kept pace with rising demand in developing regions, and the Bank has declared that about half of African countries are "...in an energy crisis." The total energy consumption in Africa is just 3% of the global consumption.

The Sun Is Rising in Africa and the Middle East: On the Road to a Solar Energy Future
Peter F. Varadi, Frank Wouters, and Allan R. Hoffman
Copyright © 2018 Pan Stanford Publishing Pte. Ltd.
ISBN 978-981-4774-89-5 (Paperback), 978-1-351-00732-0 (eBook)
www.panstanford.com

In many parts of Africa, energy is a scarce commodity. Across the continent, only a small fraction of people have access to the electrical grid, mostly those in higher income brackets. The annual per capita consumption is just over 500 kWh in Sub-Saharan Africa (SSA), where only one quarter of the population has access to electricity in any form, and SSA is the only region in the world where per capita access is falling. Contrast this with the access numbers for South Asia (50%), and Latin America and MENA (Middle East and North Africa) (80%). To date, electrical services in Africa have generally only reached wealthy, urban middle class, and commercial sectors, bypassing the region's large rural populations and urban poor. Most agriculture still relies primarily on energy provided by animals and people. Based on recent trends, over 60% of the SSA population will still lack access to electricity by 2020.

One final word about access, which many analysts identify as Africa's most important developmental need: Even in areas covered by the electrical grid, power is often unreliable. Frequent power outages cause damage to equipment and discourage investment. Together with poor transportation links, unreliable power has "...stunted the growth of domestic companies and discouraged foreign firms from setting up manufacturing plants in the continent." As summarized by Professor Akin Iwayemi of the University of Ibadan in Nigeria: "...the fundamental energy question facing Africa is providing and maintaining widespread access of the population to reliable and affordable supplies of environmentally cleaner energy to meet the requirements of rapid economic growth and improved living standards."

Low levels of electrification occur despite the fact that Africa is rich in energy resources. Unfortunately, most of these resources remain untapped. What are these resources and where are they located?

According to the African Energy Policy Research Network, Africa is third in terms of world crude oil reserves (behind the Middle East and Latin America), third in natural gas resources (behind the Middle East and Europe), and second in uranium resources (behind Australia). It is also rich in a broad range of renewable energy resources, including solar, wind, hydropower, geothermal, biomass, and ocean energy. To date, only about 7% of African hydropower resources as well as less than 1% of its geothermal resources have been tapped, and exploitation of its vast solar and wind resources is just getting under way as capital investment funds finally become available. Africa's renewable energy potential is discussed below.

South Africa, a country unique in Africa in terms of developing its energy resources, has the sixth largest coal reserves on the planet, after China, the United States, India, Russia, and Australia. As a result, most of its electricity today is produced by burning coal. It also has large renewable resources that have been explored but remained largely undeveloped so far. However, this may be changing as the world becomes more sensitive to the impacts of global warming and climate change arising from carbon emissions from the combustion of fossil fuels.

An important meeting of African energy ministers took place in Johannesburg, South Africa, in September 2011, to prepare for the then upcoming 17th Conference of the Parties of the United Nations Framework Convention on Climate Change (COP 17) to be held later that year. It produced an eloquent declaration that identified "...priorities for supporting Africa's energy development agenda in a sustainable manner." Included among these priorities was "Prioritizing clean energy: Africa is richly endowed with renewable energy resources—many of which may be developed in support of a low-carbon future for the continent. With the support of financing, technology and institutional capacity building from developed countries Africa will be able to greatly enhance its economic, social and environmental development using a diversity of clean energy sources." Several institutions have accepted the challenge of helping Africa in its development, including the African Development Bank (AFDB), the World Bank (WB), and the U.S. Agency for International Development (USAID) (see Chapter 4.3), corresponding organizations in the European Union and its member states (see Chapter 4.2), and in other developed countries. All have recognized that Africa is at a crossroads with respect to its future energy development and the resultant impact on economic development. Solutions will require distributed renewable energy generation as the only practical means of meeting rural electrification needs, boosting cross-border power trade, improving the infrastructure capabilities and management of existing electric utilities, and assistance in planning low-carbon development paths (discussed in the other chapters of this book). Putting this issue in context, H. E. Andris Piebalgs, Commissioner for Development Cooperation with the European Commission (AEEP/Africa-EU Energy Partnership), has stated: "No energy means no sustained or sustainable economic growth, no sustainable agriculture, no quality healthcare, and no decent education. In short, no energy means no development."

The resources are vast, but in most cases not yet well documented. There is a critical need for resource assessment in Africa, an essential step in developing bankable renewable energy projects. While resource assessment costs are small compared with the eventual costs of large-scale deployment, it is often overlooked early on.

Most parts of Africa receive more than 300 days of bright sunlight per year, especially the dry areas, which include the Sahara and Sahel deserts. This gives solar power the potential to bring energy to virtually any location in Africa without the need for expensive large-scale grid infrastructure. The distribution of solar resources across Africa is fairly uniform, with more than 85% of areas receiving at least 2,000 kWh per square meter per year of direct solar insolation. This is comparable to the solar insolation available in the most solar-intensive parts of the United States: New Mexico, Arizona, and California.

Morocco has one of the highest rates of solar insolation in Africa—about 3,000 hours of sunshine per year on average, but up to 3,600 hours in the desert. As a result, it has launched one of the world's largest solar energy projects, with the aim of creating 2,000 MW of solar generation capacity by 2020. Five power facilities are planned, including both solar PV and concentrating solar power (CSP) (see Chapter 2.2), which upon completion will meet almost 40% of Morocco's electrical needs. Morocco also has plans to become a net energy exporter to Europe.

Hydropower is another large renewable energy resource in Africa that is only being partially tapped. Big dams have long dominated Africa's electricity scene and account for about one quarter of the continent's total installed electric generating capacity. However, in many African countries, hydropower's share of installed capacity is much higher—more than 50% of on-grid generation in Ethiopia, the Democratic Republic of Congo, Zambia, Mozambique, and Cote d'Ivoire. As stated earlier, while only 7% of Africa's hydropower potential has been harnessed, it is estimated that the hydropower potential of the Democratic Republic of Congo alone is more than three times the continent's current power demand. The currently installed capacity is just over 20,000 MW, with another 2,400 MW under construction. More than 60,000 MW are being planned, with an estimated potential to deliver 1,750 TWh of energy. Some estimates put the added capacity in the next two decades at 80,000 MW.

African wind and wave energy resources are also large. Africa has a very long coastline, where wind power and wave power resources are abundant but poorly assessed and underutilized. While the 1,100 MW of installed wind power on the continent currently makes up only 1% of total electricity supply, at least another 10,500 MW are in the pipeline and much more is expected. Most of this activity is on Africa's western coast and Egypt.

Geothermal resources are abundant as well. They are concentrated in the area of the East African Rift, a 3,700 mile-long geological feature that stretches across 13 countries from Eritrea in the north to Mozambique in the south. With the exception of Kenya, which is one of the top 10 producers of geothermal energy in the world (530 MW installed), geothermal exploration and development has been limited. Kenya's estimated geothermal potential is about 10,000 MW. In contrast, other countries that straddle the Rift Valley have carried out no or limited exploration of geothermal resources. Ethiopia is the only other Rift Valley country to have installed geothermal power generation, currently about 70 MW.

Africa also has tremendous potential for utilization of biomass energy, its oldest and most widely used source of energy. However, this resource is also poorly assessed. The international Renewable Energy Agency (IRENA) published a literature review in 2013 (*Biomass Potential in Africa*) that concluded: "Due to the large range in results presented by the reviewed studies, no definite figures regarding the availability of biomass can be provided." This needs to change if countries are to make effective use of this abundant resource.

The opportunities for renewable energy in Africa were accurately described by IRENA Director General Adnan Amin in an article entitled "Renewable Energy Will Power Africa's

Ambitious Future": "Africa is blessed with plentiful land and natural resources. Prodigious sunshine blankets the continent for much of the year, ideal conditions for solar power. Hot rocks in areas such as the Rift Valley store geothermal energy. Vast plains and mountain ranges are great sites for wind turbines while mighty rivers like the Zambezi can be harnessed for hydropower projects. Finally, biomass is abundant—all providing multiple opportunities for renewable energy production." The promise of Africa was also recognized in a speech to the Brookings Institution by U.S. Senator Chris Coons of Delaware: "From urbanization and economic growth, to public health and energy, Africa is developing at a pace that rivals nearly every other region of the world. It is truly the continent of the 21st century."

1.1.2 The Middle East

The Middle East is a transcontinental region centered on Western Asia and Egypt in North Africa. The term MENA, an acronym for the Middle East and North Africa, has no standardized definition, and covers an extensive region reaching from Morocco to Iran. It is often used interchangeably with the term the Greater Middle East. Taking its minimal definition, the MENA Region is populated by about 380 million people, about 6% of the total global population.

The MENA region has vast reserves of petroleum and natural gas—eight of OPEC's 14 member countries are in MENA. A commonly accepted estimate is that the region has 73% of the world's proven oil reserves and 50% of the world's natural gas reserves. These reserves make it a major source of global energy, and, when combined with its location between three continents (Europe, Africa, and Asia), a region that has been in conflict since the collapse of the Ottoman Empire at the end of World War I.

The Middle East remained the world's largest oil-producing region in 2016 and accounted for all of the net growth in global production of oil (5.7%) and crude oil exports. This growth was driven by increased production in Iran, Iraq, and Saudi Arabia. Natural gas production also increased by more than 3%. The region accounted for 46% of global crude oil exports and

17% of refined product exports. Seventy-three percent of its oil exports went to the Asia-Pacific region and 13% to Europe.

Energy consumption in the Middle East, 6.7% of the global consumption in 2016, grew by 2.1% and was almost entirely provided by fossil fuels—51.5% natural gas and 46.7% oil. The share of fossil fuels in the Middle East's primary energy is the highest of any region and well above the world average of 86%. There obviously has been little incentive for the examination of alternative energy sources for its energy markets.

In 2016, renewable energy generation did increase by 42% but still constituted only a small fraction of primary energy generation. In terms of energy intensity, the amount of energy required per unit of GDP, the region continues to lead the world.

The MENA region, already a major energy consumer, together with Asia, is forecast to account for the majority of global energy demand growth over the next few decades. Historically seen as a global energy supplier, MENA's domestic energy market has been seen as marginal and well supplied with regionally produced and low-cost fossil fuels. This perception has undoubtedly fueled the region's domestic energy demand growth, led to a rapid rise in living standards, especially for the region's oil and gas exporters, and encouraged a push into energy-intensive industries.

However, this rise in regional energy consumption comes at an economic cost: rising prices for oil in recent years have raised the cost of imported oil and oil products for those MENA countries that import their energy, while the countries that are net exporters divert growing shares of their production away from international markets to meet domestic demand. This loss of revenue from diverted exports is known as an opportunity cost.

Given these new realities, and the region's large solar and wind resources, it is often argued that renewables could offer the region an important and cost-effective alternative to fossil fuels in power production. When opportunity costs are taken into account, and environmental costs associated with combustion of fossil fuels are considered when pricing fuels, the MENA region could possibly do without the renewable energy subsidies needed in other markets to stimulate market penetration of renewables. Environmental costs are often overlooked in today's energy markets because they are hard to quantify.

In recent years, there has been a major shift in the thinking of government officials in MENA with respect to energy, driven largely by the challenges of growing populations and surging demand for electricity. With large indigenous solar and wind resources, and with the costs of both technologies heading steadily downward, many governments in the region have set ambitious renewable energy goals. According to the International Energy Agency (IEA) the share of renewable energy in total power generation in the Middle East is set to increase from 2% in 2010 to 12% in 2035, and this estimate may be low. At the end of 2012, the total installed electric generating capacity in MENA was 185,000 MW, of which 10% was in renewables, mostly hydropower in Iran, Egypt, Iraq, and Morocco. Besides hydropower, wind was the most common form of renewable electricity, 1,100 MW installed primarily in Egypt, Tunisia, and Morocco.

The region, overall, is blessed with the natural resources necessary for a vibrant renewable energy sector—lots of sunshine, strong winds, and, in a few places, powerful rivers. According to the World Bank, the region receives about one quarter of all solar energy striking the earth. MENA's solar potential is larger than that of all other renewable resources combined and is even thought to be large enough to meet the current global demand for electricity, about 21,000 TWh. The MENA region also has significant wind potential, and average wind speeds in countries such as Egypt, Tunisia, and Morocco are some of the highest in the world. Many MENA countries have close to 100% access to electricity, but an estimated 28 million people lack such access, especially in rural areas. About 8 million people rely on traditional biomass for all their energy needs.

In 2016, more than 2,000 MW of new solar generating capacity was tendered in the Arab World as hundreds of billions of dollars were invested in green energy. The year 2017 is expected to surpass this level. The energy ministers from 14 Arab countries also signed a Memorandum of Understanding to establish an Arab Common Market for electricity, confirming their commitment to the development of an integrated electricity supply system for the Middle East.

The trend toward greater investment in solar energy is MENA-wide but is particularly noticeable in a few countries such as Saudi Arabia, the United Arab Emirates (UAE), Morocco, Jordan, Algeria, and Israel. In January 2017, the Saudis issued their first competitive global tender for utility-scale solar power projects. They also announced the King Salman Renewable Energy Initiative, "...the highest level of commitment to renewable energy ever seen from the Kingdom." The UAE is positioning itself as a renewable energy technology and investment hub and is home to the International Renewable Energy Agency in Abu Dhabi. The sister Emirate of Dubai announced its plans for 75% low-carbon electricity by 2050 and expects to bring 1,000 MW of solar online by 2020 to meet one quarter of the city's needs. Morocco, which currently imports 97% of its energy, is aiming for 50% renewables by 2025 and has begun an ambitious plan to use its solar resources to become a net energy exporter to Europe. Jordan imports 95% of its energy, recently inaugurated its first wind farm (117 MW), and has announced plans for more than 1,200 MW of solar. In April 2017, Algeria announced plans for three solar plants with a total generating capacity of 4,000 MW and, like Morocco, has plans to become a green energy exporter to Europe. Israel, which lacks indigenous fossil fuels and has tenuous relations with its oil-rich neighbors, is rich in solar insolation, especially in its Negev desert. This has enabled Israel to embrace solar energy and become a global leader in solar energy research and development, especially concentrating solar power. It also pioneered in the use of solar energy to power drip irrigation for agriculture.

1.2   The Role of Solar Energy in Africa and in the Middle East

Segments of the history of mankind are named after technological achievements: Stone Age, Bronze Age, and Iron Age. We should realize that we are at the beginning of a new age: the Electric Age.

Francis Bacon in the early 1600s described the phenomena that materials like amber rubbed with fur attract other objects. As the Latin word of amber is "*electrum*," he called this effect "electric." The word "electricity" was first used by Sir Thomas Browne in his work *Pseudodoxia Epidemica*, published in 1646, to describe this interesting property of amber. Only 150 years later in 1800, Alessandro Volta discovered he could produce a steady "flow of electricity" using bowls of salt solution. The next step was when Michael Faraday in 1831 invented the method to produce the flow of electricity we call today "electric current." But only in 1882 came the first main utilization of the electric current after Thomas A. Edison invented a useful electric light bulb and switched on in New York his *Pearl Street* generating station's electrical power distribution system, which provided electricity to power lights for 59 customers.

These were the beginnings of the Electric Age. The utilization of electricity quickly proceeded and today electricity has become an indispensable and defining part of our life *almost* everywhere on the earth. Where electricity is extensively used can be seen when the earth is viewed from space at night. The word "almost" is used, because even today there are large areas in the world that look dark when viewed at night from space (Fig. 1.1).

As can be seen, Africa at night is dark compared with Europe and this is also true of parts of the Middle East. Table 1.1 lists 42 African countries and indicates in each country the percentage of the population having access to electricity. Nine countries can be considered fully electrified. In 25 countries, less than 50% of the population has electricity, and among them 9 countries have only a very small percentage of their population connected to electricity sources.

Figure 1.1 Satellite view of Africa and Europe at night.[2]

Table 1.1 Access to electricity by the population in countries in Africa[3]

Country	Access to electricity (% of population) 2014 data	Entire country	Less than 90%	Less than 50%	Less than 20%	Less than 10%
Algeria	100.0	*				
Angola	32.0				*	
Belize	92.5	*				
Botswana	56.5			*		
Burkina Faso	19.2				*	

[2]https://earthobservatory.nasa.gov/NaturalHazards/view.php?id=79793.
[3]http://data.worldbank.org/indicator/EG.ELC.ACCS.ZS.

Country	Access to electricity (% of population) 2014 data	Entire country	Less than 90%	Less than 50%	Less than 20%	Less than 10%
Burundi	7.0					*
Cabo Verde	90.2	*				
Cameroon	56.8		*			
Central African Republic	12.3			*		
Chad	8.0					*
Congo Dem. Republic	13.5			*		
Congo Rep.	43.2			*		
Cote d Ivoire	61.9		*			
Egypt. Arab Rep.	99.8	*				
Eritrea	45.8			*		
Ethiopia	27.2			*		
Gabon	90.0	*				
Gambia, The	47.2			*		
Ghana	78.3		*			
Kenya	36.0			*		
Liberia	9.1					*
Libya	98.4	*				
Malawi	11.9			*		
Mali	27.3			*		
Morocco	91.6	*				
Namibia	49.6			*		
Niger	14.3				*	
Nigeria	57.7		*			
Rwanda	19.8				*	
Senegal	61.0		*			
Seychelles	99.5	*				
Sierra Leone	13.1				*	
Somalia	19.1				*	
South Africa	86.0		*			
South Sudan	4.5					*
Sudan	44.9			*		

(Continued)

Table 1.1 (*Continued*)

Country	Access to electricity (% of population) 2014 data	Entire country	Less than 90%	Less than 50%	Less than 20%	Less than 10%
Swaziland	65.0		*			
Tanzania	15.5					*
Tunesia	99.8	*				
Uganda	20.4				*	
Zambia	27.9				*	
Zimbabwe	32.3				*	
Sub-Saharan Africa	37.5					

Almost one half (590 million[4]) of the entire 1,246 billion (October 2017 data) population of Africa living in the Sub-Sahara region have no access to electricity.

On that immense territory of Sub-Saharan Africa (about three times the size of the United States), to supply those 590 million people with electricity in the traditional way, i.e., by large central power stations and connecting them by a network of power lines and wiring, is unimaginable. The reason is that most of Africa, especially Sub-Saharan Africa, suffers from triple handicap: low density of the population, long distances, and partitioning of the continent in 1884, which left Africa with a very large number of countries. This makes the use of the traditional large-scale power stations impossible because it would need the establishment of an infrastructure to transport the continuously needed fuel: coal, oil, or gas. This would only be possible near the ocean or major rivers connecting the area to the ocean. Another reason is to transport electricity by power lines through the borders of the various countries would also be very difficult to achieve.

Other traditional large-scale electric power sources not in need of continuous supply of fuel are the nuclear power and hydropower. The establishment of nuclear power stations will most likely not be permitted for safety reasons even in countries with denser population and with advanced technical capabilities, e.g., in Egypt and in South Africa. Large or even

---

[4]https://www.iea.org/publications/freepublications/publication/WEO2017 SpecialReport_EnergyAccessOutlook.pdf

smaller hydroelectric generating stations can be established in relatively few areas of Africa. The advantage is that they require no fuel. On the other hand, they are well suited in areas with higher densities of population and require an electric grid system to reach more remote customers like in Egypt. In the Sub-Sahara region, where the 510 million live without electricity, using local small hydroelectric power stations may help to solve the problem. However, large ones like the proposed 39 GW Grand Inga hydroelectric power station in the Democratic Republic of Congo (DNC) will not solve the problem, as their electricity has to be distributed by power lines that have to cross several borders, which will be very difficult to achieve and maintain.

The population of this area could be provided electricity with the help of the newly developed renewable energy systems that provide decentralized generation of electricity for individual homes and agriculture and mini-grids for small communities. Sub-Saharan Africa would need also a very large number of mostly smaller-size hydroelectric power plants which do not need transported fuel.

Small hydroelectric systems and wind turbines would be possible, but their machinery needs frequent maintenance and repairs considering the Sub-Saharan conditions. The big advantage of solar energy is that it requires no fuel as its fuel is provided by sunlight. Africa is the most sun-rich continent in the world. The biggest available energy source on the African continent is solar energy.

2

Solar Technologies for Electricity Generation

2.1   Solar Energy to Electricity: Solar cells

Life on the earth is sustained because of solar energy. Mankind used solar energy for many purposes, but the conversion of solar energy, i.e., light, to electricity was invented not even 200 years ago in 1839 by Edmond Becquerel, a French physicist. Nobody understood how the Becquerel effect—the conversion of light to electricity—worked, until Albert Einstein described it in a paper in 1905 and received the Nobel Prize for it.

For almost one and a half centuries since its invention, this effect was not useful for practically anything until 1953, when Daryl Chapin at Bell Laboratories in the United States discovered the first useful way to convert light to electricity utilizing pure silicon.

Silicon (chemical symbol "Si") is the second most abundant element in the earth's crust (27.7%). An extremely purified (99.9999999%) form of Si was used to manufacture wafers on which semiconductor[5] devices, for example, transistors and integrated circuits, could be made. Chapin and his group were able to manufacture solar cells on such Si wafers. Figure 2.1 shows the solar cell's basic structure and also how it works.

One should remember that for a battery one needs a positive (+) and a negative (–) electrode. PV battery cells operated by light also need a positive and a negative electrode.

To achieve polarity in the pure Si wafer (for solar cells, 99.9999% purity is sufficient), it has to be doped. This means that pure Si should contain an extremely small amount of an element such as boron (chemical symbol B), which will make it slightly

---

[5]Semiconductors are materials of electrical resistance that fall between conductors and insulators. There are elemental semiconductor materials that in pure state have such characteristics: germanium (Ge), silicon (Si), and selenium (Se). And there are solid solution–type semiconductors: gallium arsenide (GaAs), gallium aluminum arsenide (GaAlAs), silicon carbide (SiC), lead or cadmium telluride (PbTe or CdTe), and so on (see Wikipedia).

The Sun Is Rising in Africa and the Middle East: On the Road to a Solar Energy Future
Peter F. Varadi, Frank Wouters, and Allan R. Hoffman
Copyright © 2018 Pan Stanford Publishing Pte. Ltd.
ISBN 978-981-4774-89-5 (Paperback), 978-1-351-00732-0 (eBook)
www.panstanford.com

positively charged (p type). As shown in Fig. 2.1 (the numbers 1–4 in the following text relate to the figure), the direct conversion of sunlight to electrical energy happens in the Si wafer (1), in such a way that the absorbed sunlight directly creates electrical charges in the material. In the next step (2), on the side of B-doped Si, which will be exposed to light, a thin negative "charge separation layer" has to be formed. The thin "charge separation layer" is created by diffusing phosphorus (chemical symbol P) into the Si wafer at an elevated temperature. Once this is completed, the positive rear contact (3) is made, and on the side where the light reaches the solar cell, a grid of metal contact (4) is created, which should not block much light from the wafer.

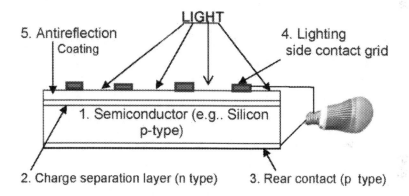

Figure 2.1 Schematic of a silicon solar cell.

When light strikes the solar cell, it develops a voltage difference between the light side and the back side. The Si base solar cell, for example, when illuminated produces approximately 0.5 V. How much power the solar cell will supply depends on the size of the wafer, its efficiency (the percentage of light converted into electrical energy), and obviously the intensity of the illumination.

There is only one step (5) in Fig. 2.1 that has not been explained yet. It is an anti-reflection coating. Its purpose is, like for photographic camera lenses or for eyeglasses, to increase the amount of light able to enter the Si material, thereby improving its efficiency.

Several utilization of these silicon solar cells (also named photovoltaic (PV) cells) were considered, but these were impractical because of their high manufacturing cost at that time. Five years later, in 1958, however, from sheer desperation, PV cells were used on the Vanguard satellite, because the batteries providing electricity for the operation of the electronics on the spacecraft, such as radio receivers/transmitters, lasted only about 20 days. Early satellites, starting with the Russian Sputnik, were operated by batteries and all failed in about 20 days— as mentioned. From desperation, on the American satellite Vanguard 1, the use of solar cells to recharge the battery of a radio transmitter was tried and compared with another similar radio transmitter powered only by a battery. Surprisingly, the radio transmitter powered by the battery recharged continuously by solar cells operated for 6.5 years, when the satellite' radio stopped working because one of its electronic circuits failed. The solar cells would have run it longer. The other radio transmitter powered only by the battery operated for only 24 days. It was realized from this experiment that satellites would have been useless without solar cells.

When it became evident that satellites and space exploration would not have been possible without high-efficiency solar cells, which were also durable in the harsh space environment, the U.S. government (NASA and the Air Force) the U.S. Communication Satellite Corporation as well as the European countries (European Space Agency—ESA) engaged in a serious effort to achieve this. Ultimately, in 1972, Joseph Lindmayer at Comsat Laboratories achieved a 15% efficient durable solar cell, called "black cells"— used in all of the satellites for at least two decades, following which newer solar cells with higher efficiency were developed utilizing other materials.

It was evident that solar cells could have great utilization for terrestrial applications, too, but for space applications, their efficiency to produce the maximum electric power on the space available on the surface of the satellite and durability to withstand the very harsh space environment were paramount. The cost of solar cells was immaterial compared with the cost of the entire satellite, its launch, and its maintenance in space. For terrestrial applications, low cost was the most important and efficiency was only secondary.

Since about 1972, the development of solar cells to convert light to electricity split into two branches. One was for space applications, achieving high efficiency and durability in the space environment. Today 35% efficient and high-durability solar cells are available; their price could be high, e.g., $100/W. The other branch was to develop solar cells and modules (a solar module is a panel in which many solar cells are connected in series; see Fig. 2.2) with efficiency of 15–20% and durability in the terrestrial environment, which is substantially less severe than the space environment, but with very much lower cost than the space solar cell.

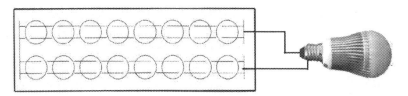

Figure 2.2 Photovoltaic module (shown nominal 6 V).

For some experts, it became evident that to reduce the prices of solar cells for terrestrial use required a new technology different from that of the solar cells used for space applications. To reduce prices significantly, mass production of solar cells, automation of their manufacture and their assembly into modules, and development of a sustainable market for the terrestrial use of the PV cells were needed.

In 1973, two companies started in the United States to achieve this. Joseph Lindmayer and Peter F. Varadi started Solarex Corporation, a private company, and Eliot Berman started Exxon's division Solar Power Corporation. Both used silicon wafers to convert light to electricity.

As predicted with the gradually increasing market, the price of silicon solar cells and modules decreased quite rapidly, and by 1978, the utilization of silicon solar modules started in Africa, too. In West Africa, French company Leroy-Somer installed silicon solar modules to power water pumps for human as well as agricultural use. Silicon solar modules were used in Egypt for navigational aids in the newly reopened Suez Canal, for pumping water in the East-Owainat area of the Sahara, and for corrosion protection of oil pipelines in Libya. Many small applications were installed in South Africa and Namibia. In the 1990s, the European

Union established a very successful large-scale program to provide water in the Sahel region in the Sub-Sahara part of the continent (Chapter 5.2 describes PV water pumping).

In the 1990s, the World Bank (WB) established programs to provide electricity for the people in regions that could not be connected to the electric grid (the World Bank PV program in Africa is described in detail in Chapter 4.4). It started to promote the utilization of PV-powered solar lanterns to replace the expensive and dangerous kerosene lights and the so-called Solar Home Systems (SHS) used to provide electricity for lights, radios, and TV systems in homes. The SHS program in Kenya started as a disaster. The problem was that many PV companies manufacturing PV modules in order to lure customers brought to the market cheap PV products. They achieved cheap modules by cutting corners in the materials used, manufacturing, and quality testing. This was especially true because some manufacturers started not to use the solar cells made on silicon wafers but manufactured a new type of solar cell using a thin film of silicon. When the World Bank realized this, adjustments were made to their program by insisting that only quality-wise approved solar lanterns and Small Home Systems (SHS) were approved to buy. With this adjustment, the program was successful not only in Africa but all over the world in many countries.

As already described, there are two different types of solar modules. One is assembled from solar cells made on a Si wafer, and the other is a thin film deposited on glass or a plastic sheet.

2.1.1 PV Modules Made of Solar Cells Created on Si Wafers

Two different types exist:

(1) Solar cells made on single-crystal wafers

From melted Si, one can "grow" large silicon single-crystal ingots. Slicing these Si single-crystal ingots, thin wafers can be produced (see Fig. 2.3).

The disadvantages of using the wafers made from Si single crystals are as follows:

- The machinery used to produce single-crystal ingots is very expensive.

- The production of single crystals by slicing is very slow and therefore many slicing machines are needed.
- The produced wafers are round, and the solar cells will not cover the entire area when assembled into modules (see Fig. 2.2); therefore, the electric output calculated on the entire area of the module will be lower. For this reason, production of wafers need an additional step, four sides are cut off of the ingots to make silicon wafers to fill the area of the module but a significant amount of the original silicon ends up as waste (Fig. 2.4).

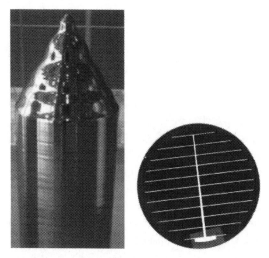

Figure 2.3 Si single-crystal ingot Solar cell produced on a Si single-crystal wafer.

Figure 2.4 Single-crystal solar cells made to optimize the electric output of modules.

(2) Solar cells made on multicrystalline Si wafer

A lot of research was conducted to develop a method to produce less expensive Si wafers for manufacturing solar cells. As a result, on November 12, 1976, at an IEEE PV conference, J. Lindmayer of U.S.-based Solarex Corporation, the world's largest PV cell and module manufacturer at that time, and H. Fischer and W. Pschunder of German Wacker Corporation, a large Si wafer manufacturer, announced a similar casting method to produce multicrystalline silicon wafers, which the two companies developed independently.

The method to produce Si wafers by casting is very simple and required less expensive machinery compared with the single-crystal manufacturing. In the process, purified Si pieces were melted under an inert atmosphere and the molten Si was poured into a ceramic container. The container was slowly cooled and the solidified Si had multiplicity of different size crystals, as shown in Fig. 2.5. The large Si block could be cut into square bricks (Fig. 2.5, left) which could then be sliced to produce wafers (Fig. 2.5, right). This multicrystalline Si wafer was obviously not suitable for use in semiconductor devices (e.g., transistors). The question was, would this material be useful for fabricating solar cells? The answer was, yes. The efficiency of the produced solar cells would be a little lower than the one made from single-crystal wafers. However, being perfectly square, multicrystalline Si wafers would fill the module area better than the single-crystal one, therefore improving its output. They were cheaper than the single-crystal ones, and by 2016 a large majority of PV cells were made using multicrystalline wafers (see Table 2.1).

Figure 2.5 (Left) Cast "multicrystalline Si" brick and block and (right) multicrystalline Si wafer.

2.1.2　Thin-Film PV Modules

Several types of thin-film PV modules were invented, but currently only three types are in commercial production: amorphous silicon (a-Si), cadmium telluride (CdTe), and the copper indium gallium (di)selenide (CIGS).

(1) Amorphous silicon (a-Si) solar modules

The first useful thin-film solar cell was invented in 1974 by Dave Carlson. It was a thin-film solar cell made of amorphous silicon (a-Si) layers, i.e., the device was made of thin films of silicon material deposited on glass that had no crystal structure. It was and still is utilized in a variety of consumer products, e.g., calculators, watches. Amorphous silicon has a relatively low efficiency (less than 10%) and is therefore used only in a few larger-scale systems.

(2) CdTe (Cadmium Telluride) solar modules

CdTe was considered to be a good material for solar cells even in the 1950s, but except for some early experimentation, it was only in 1984 that Harold McMaster started to establish a small CdTe production. He was successful in making a small production line, but it took off only in 1999 when an investment group acquired it, renamed the company to First Solar and developed mass-production machinery. First Solar became a very successful large company, basically the only manufacturer of CdTe solar modules. First Solar is one of the largest PV manufacturers in the world. In 2016, its sales were $2.9 billion and First Solar shipped 2.7 GW of solar modules.

(3) CIS and CIGS solar modules

CIS (copper indium diselenide) was invented in 1976 by Larry Kazmerski. It started to be used in 1987 when Arco Solar made a small production line and sold some CIS solar modules.

It was found that the compound copper indium gallium diselenide (CIGS), a relative of CIS, is more desirable for the production of thin-film solar cells than CIS, and most of the production now utilizes this compound. In the thin-film group, the CIGS technology has demonstrated the highest efficiency rating, high stability, little or no degradation, and excellent performance under low-light conditions.

At some point (around 2011), about 50 manufacturers of CIS or CIGS solar modules were listed. Because of the big market erosion in 2012, many of these companies were either closed or sold. There are probably 10 manufacturers left. The largest of these is Japanese Solar Frontier, which is a subsidiary of Showa Shell Sekiyu.

2.1.3 Utilization of Various PV Production Technologies (2016 Data)

Table 2.1 shows that the global production of PV in 2016 was 82.6 GW. The majority of the production was made on multi-crystalline wafers and only a small fraction was the thin-film type.

Table 2.1 Global PV production by technology—2016[6]

PV Production by technology	2016 (GW)	% used in 2016
Mono-Si	20.2	24.5
Multi-Si	57.5	69.6
Thin film	4.9 a-Si = 0.5 Cd-Te = 3.1 CI(G)S = 1.3	5.9
2016 Total global production of solar cells	82.6	100%

2.1.4 Solar PV Systems

The voltage (open-circuit voltage) produced by a crystalline solar cell is about 0.5 V. The open-circuit voltage of a thin-film cell could be higher but does not exceed 1 V. Therefore, solar modules are made of several solar cells interconnected to produce the needed voltage (e.g., 6 V, 12 V, 100 V etc.). PV arrays (systems) are made by interconnecting modules to achieve the desired voltage and power output.

Flat-plate PV arrays are interconnected modules mounted in a fixed position. The majority of PV systems are of this type. This configuration is simple and needs no maintenance.

---

[6]https://www.ise.fraunhofer.de/en/publications/studies/photovoltaics-report.html.

Tracking PV arrays. The daily electrical power output of a fixed flat-plate PV array (system) can be increased if the array is mounted on a stand that is able to track the sun so the array always faces the sun. The tracker is an electromechanical device, which is programmed to follow the sun. The tracking system can be a one-axis tracker, which follows the sun from the east to the west. In this case, the PV system is mounted at an angle where it is most efficient during winter. The other type is the two-axis tracker, which follows the sun during the day horizontally and also adjusts for the yearly motion of the sun, vertically. Compare two identical arrays: one is in a fixed position and the other is mounted, for example, on a two-axis tracker. The tracked array will outperform the fixed one. In the United States, the annual difference between fixed arrays in favor of tracked arrays, depending on the location, will be between 25% and 45%.

The decision to utilize a flat-plate or a tracking system will depend on the price difference and also on the issues related to the maintenance. Obviously, the flat-plate system is cheaper and needs no maintenance. Comparing the one-axis tracker with a two-axis one, the former is much simpler than the latter. Therefore, a large majority of the tracking systems used are the one-axis trackers. In Africa, mostly flat-plate systems are used. One-axis tracking systems are used mostly where maintenance is available.

Concentrating PV solar (CPV)[7] systems use mirrors and/or lenses to focus sunlight onto solar cells. Because the light is highly concentrated (high-concentration PV (HCPV) is 300–1000 times and low-concentration PV (LCPV) is more than 100 times the level of normal sunlight), specially designed PV cells are used to optimize performance under those conditions. In a concentrated solar cell system, the solar cells have to be at the focal point of the mirror or lens; therefore, the system has to continuously and accurately track the sun so the sun's rays are kept perpendicular to the mirror or the lens, which requires a mechanical and electronic tracking mechanism.

---

[7]*Current Status of Concentrator PV (CPV) Technology.* Maike Wiesenfarth, Simon P. Philipps, and Andreas W. Bett (Fraunhofer Institute for Solar Energy Systems ISE, Freiburg, Germany); Kelsey Horowitz and Sarah Kurtz (National Renewable Energy Laboratory NREL, Golden, Colorado, USA), https://www.ise.fraunhofer.de/content/dam/ise/de/documents/publications/studies/cpv-report-ise-nrel.pdf.

Despite the fact that they have many advantages, CPV systems are still not frequently used. By the end of 2016, a total of only 56 CPV plants were in operation around the world. Cumulative installations (already grid-connected) are only about 370 MW. The largest is an 80 MW system in China. There are only four systems in Africa and in the Middle East (see Table 2.2).

Table 2.2 CPV systems in Africa and in the Middle East

Country	City	Type	Company and country	Capacity MW	Status*
South Africa	Touwsrivier	HCPV	Soitec, France	44.0	2015
Morocco	Ouarzazate City	HCPV	Sumitomo, Japan	1.0	2014
Abu Dhabi	Masdar City	HCPV	Abengoa, Spain	1.0	2011
Saudi Arabia	Tabuk	HCPV	Soitec, France	1.1	2014
Saudi Arabia	Nofa	HCPV	Silex Systems, Australia	1.0	2014

*Refers to the year in which the project began its operation.

2.2 Concentrating Thermal Solar Power (CSP) Systems

CSP technologies use mirrors to reflect and concentrate sunlight onto an absorber where it is converted to heat. A working fluid carries this heat into a conventional thermal power plant where the heat produces steam that drives one or more turbines, which then turn generators to produce electricity. The two main types of CSP systems in use today are parabolic trough (Fig. 2.6) and solar towers (Fig. 2.7). Globally approximately 220 systems of both of these types are planned or in operation. Those in operation produce a total of 5.1 GW of electricity.

Figure 2.6 Enlarged view of the parabolic troughs CSP. California's Mojave Desert 354 MW[8] (Courtesy of NREL).

The advantage of the CSP system is that the produced heat can be stored and the system can produce electricity 24 hours a day even when there is no sunlight. The CSP system requires direct sunlight, which makes it very suitable for Africa and the Middle East.

---

[8]http://www.nrel.gov/docs/fy11osti/48895.pdf.

Parabolic trough CSP systems (Fig. 2.6) consist of many parallel rows of parabolic trough collectors that track the sun. In the focal point of these parabolic reflectors are pipes filled with liquids. The concentrated sunlight heats up these pipes and the circulating elevated-temperature liquid carries the heat that is used to produce electricity as described above. There are a total of 12 parabolic trough CSP systems in operation or in construction in Africa (Table 2.3) and 4 in the Middle East (Table 2.4). Details are in those tables.

Figure 2.7 Solar Power Tower CSP, California (Courtesy of Sandia National Laboratories).

Solar Power Tower CSP systems use sun tracking mirrors (heliostats) to focus light on the "receiver" on top of a tower (see Fig. 2.7). The receiver usually contains molten salt (sodium nitrate and potassium nitrate) as a heat storage medium. This material is usually selected because it is used in the chemical and metal industries as a heat transport fluid; so experience with molten-salt systems exists for non-solar applications. As indicated in Tables 2.3 and 2.4, there are only two solar power tower SPC planned or in operation in Africa and the Middle East.

As already mentioned, concentrating solar power systems could be highly suitable in Africa as well as in the Middle East because most of that area has abundant direct sunshine and these systems do not require the transportation of fuel. Furthermore, they could supply electricity 24 hours a day, This was realized and as Tables 2.3 and 2.4 indicate, many CSP plants have already been constructed and many are under construction in several countries of Africa and the Middle East.

Table 2.3 Operational and under-construction CSP plants in Africa

Country	City	Name	Parabolic	Tower	Capacity (MW)	Status
Algeria	Hassi R'mel	ISCC Hassi	✓		20	2011
Egypt	Kuraymat	ISCC Kuraymat	✓		20	2011
Morocco	Ait Baha	Airlight Energy	✓		3	2014
	Ain Ben Mathar	ISCC	✓		20	2010
	Ouarzazate	NOOR I	✓		160	2015
	Ouarzazate	NOOR II	✓		200	2017
	Ouarzazate	NOOR III	✓		150	2017
South Africa	Poffader	KaXu Solar I	✓		100	2015
	Groblershoop	Bokpoort	✓		55	2016
	Upington	Khi Solar One		✓	50	2016
	Poffader	Xina Solar One	✓		100	2017
	Kathu	Kathu Solar Park	✓		100	2018
	Postmasburg	Redstone Solar		✓	100	2018
	Upington	Ilanga I	✓		100	2020

Table 2.4 Operational and under-construction CSP plants in the Middle East

Country	City	Name	Parabolic	Tower	Capacity (MW)	Status
Kuwait	Kuwait City	Shagaya CSP	✓		50	2017
Saudi Arabia	Duba	ISCC Duba 1	✓		43	2017
Saudi Arabia	Waad Al Shamal	Waad Al Shamar ISCC	✓		50	2018
United Arab Emirates	Madinat Zayed	Shams 1	✓		100	2013

As it can be seen in Africa and the Middle East, parabolic trough systems (16) are more popular than solar power tower systems (2).

2.3   Hybrid Solar Systems (HSP)

In the previous two chapters, two systems were described that are capable of converting solar energy to electricity.

One is the photovoltaic (PV) "solar cell system," which uses "solar cells" (Chapter 2.1) to convert light directly to electricity. The other system is the "concentrating solar power" (CSP) system (Chapter 2.2), which converts direct sunlight to heat to produce steam. Like conventional thermal power plants, the steam is fed to a steam turbine to produce electricity. The similarity between the two systems is that both utilize the sun's energy to produce electricity and do not need expensive fuel to operate the machinery and expensive infrastructure (roads, harbors, railway systems, pipe lines, etc.) for the delivery of the electricity produced.

There are, however, differences between the two "solar" systems. CSP systems focus the solar irradiance and can therefore only use *direct sunshine*, which limits their use in the areas typically away from the coast, where humidity and clouds reduce the direct sunshine. PV converts all *light* (*not only direct sunlight*), for example, the light reflected from clouds; so its utilization is not limited. Another difference is that while PV systems produce electricity only during the daylight hours, the CSP system with integrated thermal storage can produce electricity 24 hours a day. The storage capability and technical complexity make the price of CSP electricity higher than that of PV electricity. However, between the 55 parallels north and south, solar PV is equally or less expensive by some margin than the electricity produced by the currently used conventional methods (e.g., coal, oil and nuclear), and even CSP can compete nowadays.

For example, in Dubai a contract for building a 700 MW CSP system in the Mohammed bin Rashid Al Maktoum Solar Park was awarded to the consortium of Shanghai Electric and Saudi Arabian ACWA Power[9] at an impressive bid to supply electricity for 35 years at 7.3 U.S. cents per kWh. The 700 MW system consists of three 200 MW parabolic troughs and one 100 MW power tower system. Interestingly, the CSP system is not allowed to dispatch electricity between 10 am and 4 pm, because they

---

[9]https://cleantechnica.com/2017/09/19/dubai-awards-700-mw-solar-csp-contract-mammoth-mohammed-bin-rashid-al-maktoum-solar-park/.

will have PV electricity during that time. Dubai is also in the process of building an 800 MW PV system at the same site, in addition to the already built 200 MW PV system. The Dubai Electricity and Water Authority (DEWA) received three bids to supply electricity for 20 years. The lowest solar PV price bid was 2.99 cents/kWh, and the other two were 3.65 cents/kWh and 3.95 cents/kWh.[10] Together, these systems provide solar electricity 24 hours per day at costs lower than the current gas-fired power plants.

A large part of Africa and the Middle East during a very large part of the year receives direct sunshine suitable for CSP systems. PV systems also deployed in the same area as the two systems as described above are compatible. PV systems are producing electricity during the time of the highest demand and the CSP can produce electricity at night, providing solar electricity 24 hours a day.

Similar to the Dubai philosophy, the Moroccan Agency for Solar Energy (MASEN) announced in 2016[11] a plan to install a total of 600–800 MW CSP/PV hybrid project in the NOOR Midelt solar power complex in two 400 MW phases. Each phase will consist of 150–190 MW CSP capacity and a PV component of 150–210 MW, with a total capacity of 300–400 MW.

The hybrid concept and its advantages were described in a World Bank report[12]:

> *The Project is expected to adopt an innovative hybrid solar CSP/PV plant design to combine the benefits of both technologies to deliver firm power to the grid. The feasibility study indicates that this type of hybridization is an attractive solution as it increases the load factor and reduces intermittency, compared to pure PV solutions. At the same time, it also reduces upfront investment cost and the levelized cost of energy (LCOE), compared to pure CSP solutions.*

---

[10]https://cleantechnica.com/2016/05/02/lowest-solar-price-dubai-800-mw-solar-project/.

[11]https://www.pv-magazine.com/2016/12/12/moroccos-pv-170-mw-tender-concluded-new-solar-and-energy-storage-tender-under-way/.

[12]http://documents.worldbank.org/curated/en/383431481787464986/pdf/111018-ISDS-P161893-Concept-Box396338B-PUBLIC-Disclosed-12-13-2016.pdf.

It is also expected that in the future an electrical storage system (e.g., batteries) will also be added to the HSP system to store electricity at times when the system produces surplus electricity. With proper design and computer control, this advanced HSP system would be superior to and produce cheaper electricity than the presently used coal- or gas-powered utilities, where to satisfy the peak power demand, additional inefficient "peaking" stations had to be installed, making the peak power electricity prices high.

The HSP electricity-generating stations in Africa and the Middle East, where there is abundant sunshine, will become widely used. Because they produce electricity 24 hours a day like coal or fossil fuel plants and have tremendous advantages, they do not require an expensive infrastructure for the transportation of the required fuel. Furthermore, the systems are immune to any increase and fluctuation in fuel costs, as their fuel comes directly from the sun.

3

Electric Grid Issues in Africa and the Middle East

3.1 Introduction

This section discusses some of the electrical infrastructure needs of the African continent and the Middle East region and some of the steps that are taken to address these needs. It includes (1) discussion of mini-grids, which are among the viable ways to deliver electricity services to remote, off-grid locations; (2) the efforts under way to link power pools, that help reduce costs, optimize the integration of renewable power and enable cross-border power exchange; and (3) the efforts by Arabian Gulf states to operate a multi-state grid and establish a shared power market.

The Sun Is Rising in Africa and the Middle East: On the Road to a Solar Energy Future
Peter F. Varadi, Frank Wouters, and Allan R. Hoffman
Copyright © 2018 Pan Stanford Publishing Pte. Ltd.
ISBN 978-981-4774-89-5 (Paperback), 978-1-351-00732-0 (eBook)
www.panstanford.com

3.2 Mini-grids[13]

The International Renewable Energy Agency (IRENA) describes the difference between off-grid systems and centralized grids in two ways[14]: First, off-grid systems are smaller in size and the term "off-grid" itself is very broad and simply refers to "not using or depending on electricity provided through main grids and generated by main power infrastructure." Second, off-grid systems have a (semi)-autonomous capability to satisfy electricity demand through local power generation, while centralized grids predominantly rely on centralized power stations. As opposed to stand-alone systems for individual appliances/users, mini-grids serve multiple customers. Mini-grids have been in existence for a long time but are traditionally powered by diesel generators or in some cases, small hydro installations.

According to the World Bank, mini-grids could present a least-cost and timely option for up to 400 million people in Sub-Saharan Africa and South Asia who currently do not have access to electricity.

According to IRENA,[15] Africa has around 35 GW of oil-fired generation capacity, and sub-Saharan Africa has around 19.7 GW, with significant capacity in South Africa, Sudan, Kenya, Nigeria, and Senegal. However, the average size of these utility-scale diesel generators is only around 6.3 MW each. The total market size in Africa for these generators has varied between 1.3 GW and 1.7 GW per annum. Solar PV can help displace this demand and reduce the diesel-powered generation in Africa through a combination of utility-scale deployment, solar PV mini-grids (with or without storage) and solutions such as solar home systems.

According to REN21, 2015 saw a lot of activity in the field of clean energy mini-grids in Africa.[16] One company, Powerhive,[17]

---

[13]See also Allan R. Hoffman (2016). Chapter 2.3, Grids, mini-grids, and community solar, in *Sun towards High Noon* (Peter F. Varadi), Pan Stanford Publishing, Singapore, p. 24.

[14]http://www.irena.org/DocumentDownloads/Publications/IRENA_Off-grid_Renewable_Systems_WP_2015.pdf.

[15]IRENA (2016). *Solar PV in Africa: Costs and Markets*, ISBN 978-92-95111-48-6 (PDF).

[16]http://www.ren21.net/wp-content/uploads/2016/06/GSR_2016_Full_Report_REN21.pdf.

[17]http://www.powerhive.com/.

secured a loan of $6.8 million from the U.S. Overseas Private Investment Corp. (OPIC) with a plan to build 100 solar-powered micro-grids in Kenya, which would power about 20,000 households and businesses. Enel Green Power announced that it will invest $12 million for the construction and operation of a 1 MW portfolio of mini-grids in 100 villages. The International Finance Corporation launched a $5 million program to develop a market for mini-grids in Tanzania to increase access to energy, while in Mozambique, Energias de Portugal (EDP) secured $1.95 million to finance a 160 kW hybrid solar/biomass mini-grid to power 900 households, 33 productive users, and 3 community buildings.

A scan using the interactive map on the REN21 website[18] reveals their estimate that there are now close to 150 mini-grids with solar PV in operation in Africa, which is not much given the size of the continent. One of the hurdles is the initial investment, which is substantial, and the fact that the business model is often not "bankable" in the traditional sense. Typically, the customers of the mini-grid in more rural areas are among the poorest of the poor, without any credit rating. This doesn't mean that they will not pay for good service, especially if it is better than traditional energy services based on expensive kerosene or batteries.

3.2.1 Devergy

Devergy, a company based in Tanzania and founded by two Italians Fabio De Pascale and Gianluca Cescon, has developed a very clever solution for initial sizing of the mini-grid. The core of the system is a modular tower with PV and batteries, smart electronics, and DC cabling to the customers. If in a village a sufficient number of potential customers want to be connected to the grid, a Devergy technician installs a tower and connects those customers to the grid. The customers use their own DC appliances, which are also offered by Devergy. The customers buy electricity services for a week in a prepaid setup. According to the company, the customers buy lighting for a week while they choose to power their TV for an evening. It also means a plan runs out sooner if it is used to power a fridge when it was designed

---

[18]http://www.ren21.net/resources/charts-graphs/dre-map/.

to charge a phone. So far, in 11 Tanzanian villages, Devergy has installed local mini-grids powered by solar panels and batteries. The energy consumption is remotely monitored 24/7 with innovative and unique proprietary software. When there is a need for more energy, Devergy's technical support desk is automatically informed exactly where the need exists. A technician arrives in the village to set up a new solar tower within hours to allow the grid to grow. This approach avoids the problem sometimes seen with mini-grids, i.e., that of large capacity, for which there is often not the matching demand, or capacity to pay, especially in the early years of the business. Or, sometimes the demand is larger than the initial setup can serve, and there is insufficient capacity to grow fast enough. Devergy's clever business setup, informally known as a "skinny mini-grid," keeps the capacity of their mini-grids at the minimum necessary to provide exactly the energy the customers need.

3.2.2   Donor Support for Mini-Grids

In 2016, the World Bank's Energy Sector Management Assistance Program (ESMAP) launched their Global Facility on Mini-Grids.[19] Focusing on Sub-Saharan Africa, South and East Asia, and Small Island Developing States (SIDS), the Global Facility has two main focus areas: operational scaling and global knowledge development and learning. The program cannot financially support the actual investment in a mini-grid but undertakes studies and activities that support the development of mini-grid businesses. The World Bank facility is designed to complement other programs. For example, it constitutes the knowledge component of the Green Mini-Grids Market Development Programme, which is funded by Denmark, Italy, the United Kingdom, and the United States and is part of the Sustainable Energy for All initiative.

3.2.3   Central vs. Individual Uses

Most mini-grids serve households in community settings in villages or small towns, but some also serve a commercial or

---

[19]http://pubdocs.worldbank.org/en/100101467904363149/Mini-Grids-Apr-2016-v3.pdf.

industrial user such as a mining operation in a remote area, a local business such as a printing shop or a supplier of irrigation, or a telecommunication tower. Such anchor-customer-based mini-grids, when tied to a specific economic purpose and when well maintained, have proven to accelerate economic development, as analyzed by the Rockefeller foundation in research assessing more than 106 mini-grids in India.[20]

One interesting idea relevant to the development of mini-grids is to look for so-called "killer applications" such as cooling facilities.[21] India's Ecozen is piloting six solar-powered cold rooms as anchor applications in mini-grids, which may become an important element in the so-called agricultural cold-chain close to the farmer. Often in some parts of the developing world, agricultural produce goes to waste, especially high-value crops or meat or fish, for lack of cold storage. Recently a fishing village along Mauritania's northern coast installed a series of small wind turbines that, apart from providing electricity to homes, also produce ice so the fishermen can keep their catch fresh for a longer time. It seems logical to look for a viable business opportunity beyond supplying electricity to homes, to maximize economic development as an important driver. The experience from India seems particularly relevant for Africa.

The above examples show the tremendous potential if some of the barriers can be overcome. Those barriers in some markets include limited access to finance, with the right terms and tenure, lack of entrepreneurs and skilled technicians, and the need for appropriate business models. The regulatory environment can be a serious challenge, not only in terms of the time taken to get necessary permitting to install a mini-grid and the legal gray area in terms of the development of mini-grids in some markets, but also in terms of tariffs charged. Often African countries try to support poor customers by providing them with cheap electricity, cross-subsidized by richer people in the cities and higher tariffs charged for businesses. While the intention is good, if there is no exception for mini-grids, the low tariffs in such

---

[20]www.smartpowerindia.org/documents/SmartPowerIndia_report_April_2017.pdf.
[21]https://medium.com/energy-access-india/is-cold-storage-the-next-killer-app-for-green-mini-grids-1538b12b5a31.

a setup do not enable a viable business model for mini-grids in remote areas. As a result, in absence of such a mini-grid, the people pay much more for services in the form of disposable batteries, kerosene, or diesel.

3.3   Regional Power Pools in Africa

Good infrastructure is crucial for economic development. It therefore comes as no surprise that according to the Ernst & Young 2013 Attractiveness Survey, 44% of business people in Africa listed improving infrastructure in their top two constraints to business operations, ahead of bribery and corruption. Apart from roads, railways, and (air)ports, the energy sector requires greater and smarter investments, both in generation and in transmission and distribution (T&D). The other sections of this book deal with distributed generation, which has the ability to provide energy services to people where the grid doesn't reach. In addition, countries link their power generation assets to the demand centers via a high-voltage transmission grid for the longer distances, stepping down to the distribution grids at medium and low voltage levels. Although this is inadequate in most countries—Africa has the lowest electrification rates in the world—most countries have plans to expand the generation and T&D networks to overcome the present gap and cater for growing demand. These national efforts need to be complimented by regional and, to some extent, continental initiatives for several reasons.

First, many generation projects are very large in terms of investments and would benefit from the ability to sell power abroad. Ethiopia's Renaissance dam will have a capacity of 6 GW and will cost $4.7 billion. Democratic Republic of Congo's (DRC) proposed Grand Inga Dam is even much larger with a capacity of 39 GW and an investment volume of $80 billion, twice the size of the massive Three Gorges Dam in China. No single sub-Saharan African country, apart from perhaps South Africa, can presently master such investments without involving neighboring countries. Interconnections with neighboring countries' grids are therefore crucial. Second, integrated markets reduce costs. At present, the average tariff per kWh in the region is US$ 0.14, compared with US$ 0.04 in Southeast Asia, which is better interconnected. Third, integrating electricity grids increases the availability of power, making the system more robust and secure.

Therefore, African countries started to integrate their power sector infrastructure, via regional power pools:

- The South African Power Pool (SAPP), established in 1995
- The North African power pool (Comité Maghrébin de l'Electricité COMELEC) established in 1998
- The West African Power Pool (WAPP), established in 2000
- The Central African Power Pool (CAPP), established in 2003
- and the East Africa Power Pool (EAPP), established in 2005

The power pools are initiatives to establish regional power markets and help harmonize energy policy. Nonetheless, according to the Infrastructure Consortium for Africa,[22] as far as power trade within power pools is concerned, electricity traded across border is still low in CAPP (0.2% in 2009) and in EAPP (0.4% in 2008). It is higher in COMELEC (6.2% in 2009), in SAPP (7.5% in 2010) and in WAPP (6.9% in 2010). SAPP is at a more advanced stage with 28 bilateral contracts already signed between the member countries and with an active role played by the short-term electricity market (STEM) since 2001 and by the day ahead market (DAM) since 2009. Further development of the regional market is, however, constrained by the lack of generation capacity linked with congested and insufficient interconnection capacity. But it is not just the lack of physical interconnectors. Institutional setup and market rules and regulations have already been implemented in SAPP and are being implemented in WAPP and under design in EAPP. However, CAPP and COMELEC have still to design and develop their power market institutions and rules.

Considering the above-described conundrum of the large generation projects that are too large for the host country, the following solution involving the power pools has been proposed:

Generation projects with regional dimension could be developed through a public private partnership/independent power producer (PPP/IPP) arrangement with an innovative approach, providing a minimum set of guarantees for investors and securing an acceptable level of competition between the operators of the regional market. This could lead to the following proposition, with a central role for the regional market. The off-take agreement would have two main components:

---

[22]https://www.icafrica.org/fileadmin/documents/Knowledge/Energy/ICA_RegionalPowerPools_Report.pdf.

(i) The first component would be a power purchase agreement (PPA) between the PPP/IPP and the national transmission system operators (TSOs) through the power pool for part of the generation output (for example, 50%). This would secure a minimum guaranteed revenue for the developer.

(ii) The second component for the rest of the generation output (remaining 50%) would be to establish bilateral contracts or to sell on the short-term market. This would secure a minimum level of competitiveness in the regional power market.

For any of these ideas to materialize, one must identify the bottlenecks and work on solutions for them. The following elements are crucial to operationalize the potential for cross-border trading of electricity in the power pools:

- adequately developed grid interconnections
- adequate generating and reserve capacity to meet demand of the pool
- liberalized markets, ideally privatized and unbundled (a separation of generation, transmission and distribution)
- a legal framework for cross-border electricity exchanges
- trust and mutual confidence among pool members
- regional regulation and mechanism for dispute resolution

Unfortunately, most of the sub-Saharan power pools do not yet meet most of these requirements, although many projects are under way to tackle these hurdles, including several interconnectors and regulatory frameworks, on both regional and national levels.

Renewable energy projects are important drivers for a more regional approach for electricity markets. The massive Renaissance and Grand Inga Dams can only be realized if part of the power can be exported, and the Kariba Dam on the border of Zambia and Zimbabwe has been a joint undertaking between the two countries from the outset. However, including larger shares of wind and geothermal projects, whose specific costs typical decrease with increasing scale, would also benefit from regional power markets. Furthermore, increasing shares of variable renewables such as solar and wind requires storage, additional dispatchable power, or interconnection. With the cost of solar and wind coming down rapidly, and now routinely undercutting other power sources, the point in time when this will become a necessity is not far away. And although transmission lines are capital-intensive projects, the cost of transmitting a kWh of electricity over long distances using a well-utilized transmission line is one of the cheaper options.

3.4 Gulf Cooperation Council Interconnection Authority

The Cooperation Council for the Arab States of the Gulf, colloquially known as the Gulf Cooperation Council (GCC), was established in 1981 and is a regional intergovernmental political and economic union consisting of all Arab states of the Arabian Gulf, except for Iraq. Its member states are Bahrain, Kuwait, Oman, Qatar, Saudi Arabia, and the United Arab Emirates. To foster closer cooperation in the field of trading electricity and electricity services, in 2001 the GCC member states founded the Gulf Cooperation Council Interconnection Authority (GCCIA).

3.4.1 Middle East

The Middle East is a region of extremes. While some countries are extremely wealthy, others are some of the poorest in the world. While it is the center of global oil and gas production, it is also a primary center of oil demand, driven ever higher by soaring peak power demand. And, already one of the driest and most water-scarce areas of the world, the region is expected to double its population in the next 40 years, which requires more and more electricity. The GCC countries' electricity demand is projected to grow at 6–7% per annum, which puts an increasing burden on domestically produced oil and gas. Already now, the UAE and Oman are net importers of natural gas, even though they produce large amounts. All GCC countries therefore aim to diversify their energy mix, with nuclear, coal, concentrating solar power (CSP), wind, and solar PV in the picture.

3.4.2 GCCIA

The aim of the GCCIA is to build and operate a GCC grid and initiate a power market. After its inception and preparatory phase, the first contracts for infrastructure works were awarded in 2005, worth approximately $1 billion. In 2009, the operation started and the first cross-border "trading" of electricity occurred in 2010. The GCC grid (Fig. 3.1) is largely a 400 kV backbone stretching from Kuwait to Oman and all countries are now connected.

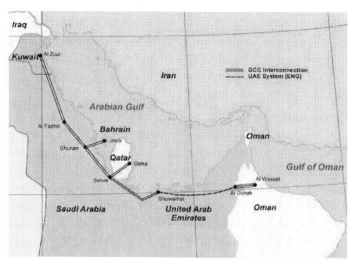

Figure 3.1 The GCC grid.

The main objectives of the GCCIA are to

- link the electrical power networks in the member states to help in emergency situations;
- reduce the electrical generation reserve of each of the member states;
- improve the economic efficiency of the electricity power systems in the member states;
- provide the basis for the exchange of electricity among the member states.

Although much has happened since its inception, one of the obstacles hindering further integration of the GCC power markets has been the high level of subsidization in the sector. All countries subsidize power, but the modalities are all different, making it difficult to find the right pricing for traded electricity. Furthermore, there are substantial differences in local regulations and sector maturity, there is no framework for harmonizing national energy policies, not every country has an independent regulator, and there are generally few market participants. Nonetheless, the GCC grid was successfully used in more than 1,300 "incidents" until the end of 2016, with a utilization factor of 8%. This has prevented blackouts, something that many GCC residents had to get used to in the 2000s. In 2016, the first commercial agreement between Saudi Arabia and Abu Dhabi (UAE) for 300 MW capacity became operational, and for the first time, cross-border electricity was supplied and paid for. All the "incidents" are usually not compensated for in monetary terms. According to calculations by the GCCIA, in the present setup the economic value that can be attained through trading could amount to $1.8 billion until 2034. If the use of the GCC grid is maximized, the total theoretical benefit of economic electricity exchanges between Saudi Arabia and Abu Dhabi could amount to $2.9 billion during the period 2012–2030 considering a link utilization factor of 50%. Similar relationships could yield benefits of $2.0 billion between Saudi Arabia and Bahrain, $720 million between Saudi Arabia and Kuwait, and $3.4 billion between Saudi Arabia and Qatar. Overall, the regional market could be worth $27 billion over 25 years.

3.4.3 GCCIA and Renewable Energy

There is very little natural sweet water in the GCC, most of the water is desalinated seawater from the Arabian Gulf. It is produced in fossil fuel power plants using the waste heat from the turbine-generators in either a process called multi-stage flash (MSF) or multi-effect distillation (MED). The electricity demand is seasonal and dips in the winter due to reduced need for air-conditioning, but the water demand is relatively constant throughout the year. This leads to low efficiencies in the winter period because the power plants are mainly run to produce water. With large amounts of electricity from solar PV and nuclear power plants, that do not directly produce desalinated water, coming online in the next few years, this situation is likely to get worse. Abu Dhabi is constructing four nuclear power blocks near the border with Saudi Arabia, with an overall capacity of 5.6 GW, and Saudi Arabia also has plans for nuclear power projects. Dubai will add 5 GW of solar power to their grid until 2030, Abu Dhabi recently launched a 1.2 GW solar PV project, and Saudi Arabia recently launched their Vision 2030, which has a 9.5 GW target for renewable energy, mostly solar.

A few solutions are key to solve this conundrum. First, decoupling of the water and the electricity sector is required. This can be achieved by producing water using reverse osmosis (RO). RO was previously costly, and not feasible due to the high salinity of the Gulf. However, recent technological developments have made RO feasible and affordable, so there are many projects in the pipeline and several Gulf countries are working on independent water producers' schemes. Second, increased interconnection through the GGC grid will enable transferring power across borders, linking generation capacity with large industrial consumers more effectively. There is growing recognition that a greater role for the GCCIA and introducing a real regional energy market would lead to substantial cost savings and overall system optimization.

4

Regional and International Solar Initiatives

4.1 Introduction

This chapter starts with a discussion of the help that the European Union (EU), the United States, and the World Bank have provided to Africa and the Middle East to advance their energy development. It includes a first-hand account of the EU efforts by Wolfgang Palz, the head of EU renewable energy programs for many years and the originator of many of these assistance programs. Assistance by the United States has been provided largely by the U.S. Agency for International Development (USAID),[23] which has primary responsibility for U.S. non-military International development assistance activities. The World Bank efforts are described by Anil Cabraal, who initiated many of the Bank's solar energy assistance programs while working at the Bank.

The rest of the chapter is devoted to a discussion of the Africa Clean Energy Corridor, a pan-African electricity sharing network stretching from Cairo to Capetown, the Global Energy Transfer Feed-in Tariff (GET FiT) program, conceived and established in 2016 by the German Deutsche and KfW Banks promoting renewable energy investments in developing regions, and a discussion of the Desertec concept, which saw an opportunity to tap into Africa's substantial solar resources to provide clean electricity for African development and export to Europe.

---

[23]A detailed description of the U.S. Government's winding road for renewable energy: Allan R. Hoffman (2016). *The U.S. Government & Renewable Energy: A Winding Road*, Pan Stanford Publishing, Singapore.

The Sun Is Rising in Africa and the Middle East: On the Road to a Solar Energy Future
Peter F. Varadi, Frank Wouters, and Allan R. Hoffman
Copyright © 2018 Pan Stanford Publishing Pte. Ltd.
ISBN 978-981-4774-89-5 (Paperback), 978-1-351-00732-0 (eBook)
www.panstanford.com

4.2   Introduction to the European Development Aid: A Personal Recollection

Wolfgang Palz[24]

The global yearly budget for development aid in 2014 amounted to $135.2 billion. Trend is increasing: It was that year 66% higher than in 2000. The budget included donations, credits, and also military aid. Leading providers of aid were the United States, the United Kingdom, Germany, France, and Japan.

The world's greatest donor was the European Union, with €55 billion. This budget was shared between the individual EU Member Countries and the EU Commission in Brussels. The EU Commission implements the European Development Fund (FED) since 1958. Currently, the beneficiaries are the 79 ACP (African, Caribbean, and Pacific Group of States) countries in Africa, and some islands and other states in the Caribbean and the Pacific. For the period 2013–2020, the current FED amounts to €30.5 billion. For implementation, the EU Commission disposes of some 1,000 officials and other staff. It operates from Brussels, mainly in the service EuropeAid and in the beneficiary countries via its representations there. The EU representatives have there an effective, even though unofficial, role of co-ordination with the embassies of the EU countries providing local aid. Once in a year, the EU representative in each country comes with his chiefs of staff for reporting to Brussels. This was at least the case when I was involved some years ago when I represented EuropeAid for 1 year in those meetings.

The official objectives of the donors' aid are, in pursuing the UN Millennium goals, the sustainable development, eradication of poverty, and primary education for everybody. A major problem for implementation is widespread corruption. So, the EU Commission adopted early on the principle of "zero tolerance" for this scourge.

Given that the lack of energy is one of the big problems in the ACP countries, one would have thought that the supply of energy was an absolute priority in the aid programs. But it was not. The renewables were not implemented on a major scale. But the conventional energies did not do any better. Perhaps after

---

[24]Wolfgang Palz's biography is on page 241.

all we escaped a disastrous development as the budgets involved would have been enough to cover the whole continent of Africa with nuclear power. As far as one can know, infrastructure development like road building got much of the support of the donors.

The Sahel countries' area represents a major part of the African continent concerned with the aid programs. The obvious problem there is the lack of sufficient water for drinking, cattle feeding, and irrigation. And decentralized solar electricity is the right option to tackle the problem. This became clear already in the 1970s. One of the first to set up solar PV pumps for the villagers in those early days was the White Father Verspieren. He installed dozens of them.

I was until my start in Brussels in 1977 an official in Paris, France. As I was one of the few to know about solar PV in the early days, I was invited to several study tours across West Africa by the UN in New York. I also became in charge of a report about the potential of solar energy for the West African Development Bank in Lomé. I teamed up with the chief official for innovation at the Ministry of Industry in Paris for touring the countries of interest. We proposed for immediate action a comprehensive program for spreading PV in the Sahel countries. The financing scheme we proposed was similar to the FIT that was adopted years later for the German market. But our report had no follow-up.

Eventually, in the 1980s, the EU Commission started to embark on the Regional Solar Program PRS. Beneficiaries were nine ACP countries in the Sahel zone—all "French speaking." The purpose was to bring water and electricity via solar PV to the rural populations. The total budget of €112 million was entirely provided by the EU. The program went from 1990 to 2007 and brought in over 1,000 installations fresh water to a population of 4 million people.

The idea for the PRS went back to 1986. That year, we organized a major European PV Conference in Seville, Spain. It was part of the EU-PVSEC series that I had started a few years earlier—a series of conferences that still brings together nowadays thousands of experts in the field every year. I had invited to Seville that year a colleague from Brussels who was directly involved in the ACP programs. It was the occasion to jointly agree on that major solar action of PRS to the benefit of the poor in Africa. Later in the preparatory process, we jointly defended the new program in the competent advisory committees. The rule for the Commission's programs involving major budgets was that they have to pass various committees with experts and national officials of the EU member countries.

PRS 1 ran from 1990 to 1998. A total of 610 PV systems were installed for drinking water, 16 systems for irrigation together with 650 lightings, and a few systems for refrigeration. After completion, the whole program was subject of detailed assessments and reviews. As the conclusions were encouraging, a follow-up program, PRS 2, was decided. In the late 1990s, I moved to the ACP program with responsibility for renewable energies and supported the decision process. PRS 2 started in 2001 and served to consolidate the installations built previously and start accompanying measures around them. Five hundred additional systems were installed as well.

In the early 1990s, on behalf of the EU Commission's development program of which I was in charge at that time, I came up with the concept "POWER FOR THE WORLD." The idea was to set up a global PV action plan in favor of the more than a billion poor who were still left aside in the worldwide spread of cheap electricity supply. Details of the concept can be found in the book with the same title that I published later.[25]

As the concept "Power for the World" called for billions of dollars and euros in a new effort for the poor, one could imagine that the political decision process was tremendously difficult and outside any reach. But it raised the question for concrete action.

What actually happened not long after I had promoted the concept was that Germany started a large-scale promotion of PV—not for the poor but for its own population. And Germany was quickly followed by Spain, Japan, the United States, the United Kingdom, and in particular China, which became eventually the world leader for bringing PV into the mainstream of global energy supply. Since the beginning of the century over a trillion of US dollars have been spent on PV and over 350 GW installed.

---

[25]Wolfgang Palz, with contributions from 41 international pioneers (2011). *Power for the World*, Pan Stanford Publishing, Singapore.

In this global move the world's poor had again been left aside. I called it in my publications of the time "the 1% scandal." Indeed, 99% of global PV production went and still goes to the "Northern countries" and very little is left for Africa and the other regions in survival needs.

But the little still makes an impact! As the cost of a PV module plunged to a fraction of a dollar a watt on the world markets, Africa does benefit as well for meeting some basic needs of the poor with PV. So, nowadays, for a few dollars you can get there a PV-powered LED lamp with a battery charger for a cellular phone or other small electric devices. For the past, one could mention a reference of the United Nations Development Program (UNDP) in 1993 for a PV lighting system in Kenya that cost $1,378 installed. Leaving alone that the price included duties and VAT, it was further inflated by the fact that low-energy-consumption LEDs were not yet available and one needed a 50 W panel at a price of $7/W to power one lamp.

Millions of cheap PV lamps are widespread nowadays in Africa. But the fact remains that half of the more than 1 billion people who still lack electricity today live on the "black continent."

Early in the new century, I served as an advisor on the EU Commission's aid program for Latin America. It is not part of the ACP. I initiated then a new regional program for spreading renewables, i.e., PV with some mini wind power, in the schools in the poorest areas of Central America—600 not yet electrified villages in eight countries. Details of this program, called Euro-Solar, which is completed since 2011, are given in a new book—together with other information.[26]

I actually returned to Africa in 2011 on behalf of the German government. I came to French-speaking Bénin and its capital, Cotonou, gave a presentation at the German embassy, talked on local TV and discussed with the members of an important solar faculty at the university. Then I went to Nigeria, Africa's largest country. It was shortly before Boko Haram started its terrorist activities there. However, in Lagos, all living quarters downtown were already heavily guarded. And in the capital, Abuja, one could not enter the government quarters; all social life of the expats was limited to just one hotel. I gave speeches at the

---

[26]Wolfgang Palz, *The Triumph of the Sun*, to be published in 2018 by Pan Stanford Publishing, Singapore.

German consulates and the embassy and met the German NGOs. I was most impressed by Kano in the Sahelian area in the north of the country. It used to be the capital of an empire and there is still an important city wall. The city has an emir with all the notables of the city—impressive with their wonderful colorful turbans. I was well received at the Goethe Institute there. It has its own solar passive building fully provided with PV electricity. I wrote in my report, isn't it a scandal, just one building equipped with PV in a country twice as big as Germany—while Germany has more than a million PV houses. I was satisfied to be able to meet the governor, who promised after hearing my speech to do a lot more on solar energy in his region. I also met local entrepreneurs in PV (Fig. 4.1). These local people are real professionals who travelled for their business as far as China.

Figure 4.1 At the workshop of a local PV entrepreneur. The author (right) and the director of the local Goethe Institute (second from the left).

I want to conclude with the remark that the recent climate initiatives with, among others, a "clean energy program" for Africa turned out to be more controversial than one would have thought. In fact, in March 2017, 19 projects of the Africa Renewable Energy Initiative (AREI) were approved—for an impressive budget of $5.2 billion. However, the decision led to a certain scandal as the African head of AREI, Youba Sokona, resigned from his job in protest. I don't want to report on this regrettable dispute as I was not involved in any way in the proceedings. I only noticed complaints that I often heard before, such as "the EU-chosen projects focus on big infrastructure rather than community-led solutions. Some aid fossil fuel generation, some will be owned by EU companies."

4.3   U.S. Energy Development Assistance to Africa and the Middle East

4.3.1   Africa

The U.S. Agency for International Development (USAID) is the lead U.S. government agency focused on development efforts to reduce global poverty. For many years, its primary focus was on providing access to clean water in African nations, but in recent years, it has put increased effort into providing assistance with energy development. This follows from the recognition that access to electricity is the biggest obstacle to realizing the full potential of economic development in Africa.

A major step was President Obama's launch in 2013 of the Power Africa initiative, which set a goal to double the access to electricity in sub-Saharan Africa. Specifically, its initial goal was to add 10,000 MW of generating capacity and supply electricity to 20 million households within 5 years. In 2014, at the U.S.–Africa Leaders Forum, President Obama renewed his commitment to the initiative, pledged $300 million per year in assistance, and set new targets of 30,000 MW and at least 60 million household and business connections. Joining the United States in this effort were several governments (European Union, the United Kingdom, Norway, Sweden) and development banks (World Bank Group, African Development Bank).

These initial efforts were followed by passage of the Electrify Africa Act of 2015, which set a legislative goal, supported by both major U.S. political parties, to provide access to electricity to at least 50 million people and to add at least 20,000 MW of power across sub-Saharan Africa. This legislation helped to assure U.S. partners that the U.S. commitment would extend beyond the end of the Obama Administration on January 20, 2017. The initiatives under Electrify Africa are coordinated by USAID, with strong interagency participation and support.

An important feature of Power Africa is that it does not just focus on providing financial aid to Africa's poor, a characteristic of earlier U.S. aid programs. Instead, it focuses on the empowerment of the African people and providing them the skills and resources they need to develop sustainable and growing economies. The paradigm shift is that Obama was the first U.S. President to treat Africa as a continent with significant potential and not just "...a place that needs rescuing."

Another important characteristic of Power Africa is its dependence on public–private partnerships to carry out its transaction-centered approach to addressing the barriers to development. This means that Power Africa works with other federal agencies and the private sector to build local capacities that help make assistance programs in Africa more effective and sustainable. These partnerships enable Power Africa to establish interagency teams focused on integrated power and transmission projects supported by financing, insurance, technical assistance, and grants from across the U.S. government and private sector partners. It also utilizes field-based professionals with energy and investment experience to serve as advisors to African governments and to Power Africa staff in each partner country.

A critical aspect of any development project is reaching closure on the needed finance. Often the hardest part of a project is securing the required funds and finalizing the necessary agreements with a host government. Excessive time to closure is often cited as a major barrier to investment by private sector developers that have to put their own resources on the line. Recognizing this, Power Africa has put great emphasis on ensuring rapid financial closure by focusing on project bankability and project and financing risk management. It also works with African governments to ensure that the regulatory environment provides a supportive environment for investment.

As discussed elsewhere in this book, decentralized energy solutions utilizing renewable energy technologies such as solar and wind and biomass are the most likely to bring electricity to Africa's remote locations. Power Africa programs support the planning, financing, and deployment of these and other energy technologies (geothermal, natural gas) to modernize Africa's energy sector. To date, as reported in its 2017 Annual Report, it has mobilized more than $54 billion in commitments from more than 140 public and private sector partners and helped close financial arrangements for 80 projects that are or soon will be generating more than 7,200 MW of electrical power (of this, approximately 10% is in solar). Its efforts have enabled private sector companies and utilities to provide power to more than 10 million homes and businesses. Specific examples include support for the government of Nigeria to finalize power purchase agreements for 14 utility-scale independent power producer projects totaling 1,125 MW. In Uganda, Power Africa is supporting four pilot projects for solar, biomass, and mini-hydro. For projects in other countries, including Ghana, Kenya, and South Africa, $1 billion in new loans and financing have been announced.

Nevertheless, Power Africa has been criticized for falling short of its goals. According to John Rice, vice chairman of General Electric (GE), in comments at the 2016 World Economic Forum in Rwanda, Power Africa has so far not had much of an impact: "...if you look today at the number of megawatts that are actually on the grid directly related to the Power Africa initiative, it is very little." Power Africa's response, from its coordinator at USAID, Andrew Herscowitz, is that "Results will take years. You can't just wave a magic wand and have the infrastructure appear—it takes time to build things." He points out that Power Africa is supporting projects across Africa, that markets for solar PV are emerging rapidly, and that the private sector is getting increasingly involved and betting tens of billions of dollars on dozens of projects. President Obama also responded by saying: "Three years after launching Power Africa we're seeing real progress." Power Africa also has Congressional support, suggesting it could survive under the Trump Administration. Republican Bob Corker sponsored the Electrify Africa Act of 2015 in the U.S. Senate, and Representative Ed Royce led its passage in the House.

Nevertheless, Power Africa's status in the early part of the Trump Administration remains uncertain. Critical appointments, such as assistant secretary of state and principal deputy assistant secretary of state for African affairs, are still to be made. The Trump campaign made no mention of the program in its campaign literature.

However, at a recent meeting of the Africa Energy Forum in Copenhagen, Andrew Herscowitz directly addressed the question of where Power Africa is headed under the new administration: "The truth is that Power Africa will continue working towards facilitating power projects in Africa, although we will now have an increased focus to boost business opportunities in the USA, e.g. creating export opportunities for U.S. companies that want to invest in Africa."

As some analysts have pointed out, Congress has good reasons to continue to support Power Africa: It leverages private investment; it earns a positive return for U.S. taxpayers (the modest grant awards are used mostly to bring private power projects to completion), and it creates jobs by creating opportunities for American companies. It is already paying diplomatic and humanitarian dividends, and over time, increased electrification of Africa will enhance American security. These will be important considerations for the Trump team, once it is in place, to take into account as it decides whether to provide continued support for Power Africa.

4.3.2   Middle East

U.S. government foreign aid to the Middle East has, in recent decades, been dominated by military spending. Two nations, Israel and Egypt, have been the major recipients since the Camp David Peace Accords were signed in 1979. Since 1946, the U.S. Congressional Research Service estimates that the United States has provided approximately $290 billion to the region. In FY2016, the aid to the Middle East accounted for one-third of the geographically specific assistance in the State Department's International Affairs budget request.

USAID also provides non-military assistance to the region. It operates bilateral missions in five countries (Egypt, Iraq, Jordan, Lebanon, Morocco), a mission in West Bank/Gaza, and an office in Tunisia. It also supports development activities in three countries without missions: Syria, Libya, and Yemen. The primary focus has been on educational programs, job training, democratic governance, and help with provision of water services. The region has some of the most water-scarce countries in the world.

Energy assistance to the region has been limited, given the region's large fossil fuel resources, and the growing understanding in recent years that the United States is less dependent on the Middle East oil. The large amounts of fossil fuels made available by fracking of shale deposits has allowed the

United States to become the world's largest producer of oil and gas and potentially a net energy exporter, and reoriented global energy markets.

Nevertheless, USAID has continued its mission of helping countries across all aspects of the energy sector to build strong systems that can power global economic and social development. Specific USAID projects have assisted Egypt and Syria in introducing solar-powered submersible water pumps for irrigation. The United States–Israel Energy Cooperation Act and follow-up acts authorized the U.S. Department of Energy (DOE) to establish a joint U.S.–Israeli grant program to fund research in solar, biomass, wind energy, and other issues. While no additional funds to support the legislation were appropriated, the authorization Act did authorize the Secretary of Energy to provide funding for the grant program as needed for a seven-year period starting in December 2007. It was reauthorized as the United States–Israel Strategic Partnership Act in December 2014 for an additional 10 years. To date, 32 joint projects have been approved, with the United States contributing $14 million and Israel $8 million. The joint program is known as BIRD (the Binational Industrial Research and Development Energy Program). (*Note*: While these numbers are significant, it is important to put them in context: The Emirate of Abu Dhabi, for example, is spending hundreds of millions of dollars on renewable energy projects in Jordan, Mauritania, Egypt, Sierra Leone, and other countries.)

In addition, efforts are under way in Congress to establish a joint U.S.–Israel Energy Center, as called for in the 2014 Strategic Partnership Act. Future action on a September 2016 DOE proposal to do so is subject to appropriations.

4.4 Lighting Africa: Evolution of World Bank Support for Solar in Africa

Anil Cabraal[27]

4.4.1 In the Beginning

The World Bank Group's foray into solar photovoltaics began in Asia. In the 1990s, increasing electricity access was a priority in Asia.[28] In Asia (excluding high-income countries), there were about 815 million rural people without access to electricity in 1992; by 2014, the number had dropped substantially to 378 million.[29] These countries were seeking World Bank support for rapidly increasing electricity access. It was a special challenge in sparsely populated rural areas, as extending the grid and providing supporting generation was not only expensive and time consuming but also non-economic as the loads to be served were small. Photovoltaics seemed an ideal candidate as it was scalable, did not need wires to deliver electricity, and could potentially offer a low-maintenance way to provide basic electricity services. It was economic when compared with grid extension, though not cheap in a financial sense. With the right set of interventions, solar PV could be a solution to bring basic electricity services to many of those who could not be served by the grid.

At about the same time, the World Bank set up the Asia Alternative Energy Unit (ASTAE). Loretta Schaeffer led ASTAE in its formative days. She was a seasoned World Bank manager who knew how to smartly navigate the World Bank bureaucracy. She recruited technical specialists in renewable energy and energy efficiency. I led the work in renewable energy. ASTAE was the catalyst for the World Bank's intervention into photovoltaic electrification. One of the first investigations ASTAE conducted

---

[27]Anil Cabraal's biography is on page 241.

[28]The World Group comprises The World Bank, which lends to developing country governments; the International Finance Corporation (IFC), which lends to the private sector in developing countries; and the Multilateral Investment Guarantee Agency (MIGA), which protects investments against-non-commercial risks. For more details, see http://www.worldbank.org/, http://www.ifc.org/, and https://www.miga.org/.

[29]The World Bank, Data Bank, http://databank.worldbank.org/data/reports.aspx?source=2&series=SP.RUR.TOTL&country=#.

was a review of experiences in Honduras, Indonesia, the Philippines, and Sri Lanka, which had begun commercially oriented solar home systems (SHS) sales. The commercial off-grid solar business was emerging at that time. Costs were high in the 1990s. The SHS being sold in these countries ranged in capacity from about 6 Wp to 100 Wp, and costs ranged $8–26 per Wp in nominal terms (in 2015 U.S. dollars, the costs would be $12–41 per Wp). In contrast, in Bangladesh today, where there is a mature market with 4.2 million SHS installed, costs for systems ranging 10–135 Wp in size are $4–8 per Wp, installed, including a 5-year system-wide warranty.

In the 1990s, product quality and performance was uncertain. Most components, other than PV modules, were made locally. Batteries were usually automotive batteries, at best modified with larger reservoirs for battery acid. Inefficient incandescent lamps were used, though fluorescent tube lights and then compact fluorescent lamps (CFLs) began to be used. Beyond the International Electrotechnical Commission (IEC) specifications for PV modules, there were none that were used for controllers, batteries, or lamps.

Out of these field investigations emerged the 1996 World Bank's guidance on design of solar electrification programs— Best Practices for Photovoltaic Household Electrification Programs.[30] The principal guidance it provided has withstood the test of time. It noted the need for accessing financing to deepen the market beyond high-income customers in the community. Lowering transaction cost, including streamlined procedures, and having banking facilities close to the borrower increased the effectiveness of financing reaching customers. Tailoring repayments to the cash flows of customers such as farmers and herdsmen was important. It advised that judicious use of grants and subsidies should be limited to market-conditioning activities or limited injections of equity to buy down capital costs. Today, with the advent of mobile phone payment schemes and technology to offer Pay-as-you-go services, credit risk and costs of financial transactions can be lowered significantly.

The report advised that it was better to rely on local capabilities for implementation, and offer training programs

---

[30]Authored by Anil Cabraal, Loretta Schaeffer, and Mac Cosgrove Davies. World Bank Technical Paper No. 324, Asia Technical Department Series, August 1996.

for technicians and users. From a rural electrification planning perspective, it recommended that off-grid SHS be part of a portfolio of technologies that can provide least-cost electricity services. Today, in countries as diverse as Rwanda and Myanmar, comprehensive rural electrification planning supported by the World Bank considers grid extension, SHS, and mini-grids in delivering least-cost electricity services.

The report emphasized that successful solar PV electrification requires satisfied customers. Success requires satisfactory performance, guaranteed by quality products as well as attention to battery replacement/recycling and customer education. It recommended adopting internationally recognized standards and enforcement. Importantly, it noted that customers of limited means should be offered smaller systems rather than compromising quality in the interest of reducing costs. In the early SHS projects funded by the World Bank, technical standards, quality testing procedures, and assistance in quality certification were supported by the World Bank, including in Indonesia, Sri Lanka, Bangladesh, and China.

It was this observation that quality was a concern and that the World Bank had to step in to develop standards and support quality testing that led Peter F. Varadi to contact me with his idea of creating the Photovoltaic Global Approval Program (PV-GAP). He rightly concluded that the industry must ensure that their own markets are not spoiled by poor-quality products. PV-GAP, for which I served as a board member, developed technical standards and quality assurance processes, including a quality mark. The World Bank supported some of its work, including preparing technical standards, and developing quality processes for SHS designers, manufacturers, testing laboratories, and technician training.[31] The PV-GAP license for its quality mark

---

[31]Four training manuals were produced: (1) *Quality Management in Photovoltaics: Quality Control Training Manual for Manufacturers*, by Peter F. Varadi, Ramón Dominguez, and Deborah McGlauflin. The purpose of this training manual is to establish global quality standards and to begin to help small companies in developing countries meet them. (2) *Training Manual for Quality Improvement of Photovoltaic Testing Laboratories in Developing Countries*, by Gobind H. Atmaram and James D. Roland (produced by the Florida Solar Energy Center). This training manual is specific to quality improvement of PV testing laboratories. It addresses the testing of PV components and small systems. (3) *Certification for the PV Installation and Maintenance Practitioner: Manual for Implementing Qualified Certification Programs*,

was later transferred to the IEC System of Conformity Assessment Schemes for Electrotechnical Equipment and Components (IECEE).

4.4.2 Evolution

By 2000, there were 12 World Bank projects or programs supporting SHS electrification—the majority in Asia (India, Indonesia, Sri Lanka, China, Lao PDR). There were only two projects in Sub-Saharan Africa in Benin and Togo. The Global Environment Facility (GEF is an international organization set up in 1992 at the Rio Summit, to address global environmental issues), committed US$ 1.1 million with US$ 4.6 million in World Bank and other co-financing for each country to fund Decentralized Rural Energy Projects. The goal was to establish financially viable private-sector installation and service companies. Although funding was approved by the GEF in 1998, it was never approved by the World Bank.

The International Finance Corporation (IFC), which is the private sector lending arm of the World Bank, experiences with off-grid solar were less positive than those at the World Bank. The IFC reviewed their early experiences with SHS and in 2007 issued a report, *Selling Solar: Lessons from More than a Decade of IFC's Experience.*[32] The report reviewed five solar PV programs that were primarily financed by the GEF with four being multi-country programs and one a grid-tied project in the Philippines. The programs resulted in over 84,000 SHS being sold.[33] However, these programs were financially

---

by Mark C. Fitzgerald (produced by the Institute for Sustainable Power, Inc.). This manual is a guide and resource for use in implementing a national quality trainer accreditation and practitioner certification system for PV systems design, installation, and maintenance. (4) *Solar Home Systems: Manual for the Design and Modification of Solar Home System Components*, by Mark R. Vervaart and Frans D. J. Nieuwenhout (produced by the Netherlands Energy Research Foundation). This manual forms the basis of a training course for engineers who are in the design and modification of solar home system equipment. http://siteresources.worldbank.org/EXTEAPASTAE/Resources/QUAP_PV_Overview.pdf.

[32]IFC, Selling Solar, http://www.ifc.org/wps/wcm/connect/topics_ext_content/ifc_external_corporate_site/ifc+sustainability/learning+and+adapting/knowledge+products/publications/publications_loe_sellingsolar__wci__1319577385747.

[33]For comparison, the highly successful Bangladesh RERED project at its peak was installing on a commercial basis, about 100,000 SHS a month.

unsuccessful and did not transform the market or create sustainable businesses. The IFC acknowledged that the problem lay not in the technology but in market realities defying expectations—price of PV did not decline, there was a shortage of smaller sized modules, and there were several economic shocks. It concluded that "supporting the growth of the solar PV market is far more complex than first envisioned, particularly due to the level of market segmentation that exists." The IFC decided to diversify from PV as a means for rural electrification and support a broader array of technologies, the commercialization of low-power lighting devices, and distributed power generation. This decision is important considering the future priority IFC accorded to Lighting Africa.

In contrast, the World Bank continued to be cautiously optimistic about SHS as a viable off-grid electrification option. By the mid-2000s, the World Bank was preparing or implementing 36 SHS projects worldwide to benefit 3.6 million households with total project investments exceeding $1 billion. Among these were projects in 12 Sub-Saharan African countries to benefit 230,000 households with a total investment of about $155 million. The global electrification goal has now been exceeded, with World Bank–assisted Bangladesh SHS projects dwarfing the other programs with SHS serving over 4.2 million households—17% of all electrified households in Bangladesh.

While the Best Practices report provided relevant advice, it was difficult to meet all the requirements in Africa. Africa is a vast continent with sparsely populated rural areas with difficult transport conditions. There was limited access to financing facilities in rural areas, and even where financing was available, it was primarily for short-term loans for income-generating purposes. Many customers were subsistence farmers or herdsmen. Higher-income customers were few and far between. There were high import duties making solar products even more expensive.

In 2008 and 2010, the World Bank issued two guidance notes to staff that updated and improved upon the best practice guidance issued over 10 years earlier. *Designing Sustainable Off-Grid Rural Electrification Projects: Principles and Practices*, issued in 2008, provided the World Bank staff and others interested in off-grid electrification with useful guidelines for designing sustainable off-grid projects.[34] The combination of high cost of service, poor customers, and newer, less familiar technology options often makes it a more complex task than preparing a conventional energy project. Higher costs and complexities of serving off-grid communities, the benefits in terms of improved living standards and opportunities income generation, for far outweigh the costs.[35] Given the unique features of projects and country situations, the note did not prescribe solutions; rather, it provided basic design principles and good practice. One important insight that was not noted previously was the need for early community consultations and gaining community confidence. This is particularly important as electrification is politically popular and some communities viewed solar PV as not being "real" electricity, even though it offered the opportunity of obtaining electricity services far earlier than waiting for the grid to reach a community.

Photovoltaics for Community Service Facilities: Guidance for Sustainability, issued in 2010, provided guidance on using solar PV to meet the electricity needs in social-sector projects with off-grid community facilities, including health and education.[36] It offered good practice guidance in design and proper care of the PV systems to assure reliable and long-term operation.

Over the past 10 years, with steady decline in PV costs, increased economic growth in many African countries, maturation of the African market infrastructures, and the emergence of three disruptive technologies—solar photovoltaics, LED lighting, and mobile phone/mobile pay—solar electrification has emerged as a dominant option in bringing electricity services to rural Africa far from the grid, at a faster pace, at lower costs, and with less dependence on governments. Several companies are

---

[34]Ernesto Terrado, Anil Cabraal, and Ishani Mukherjee, *Designing Sustainable Off-Grid Rural Electrification Projects: Principles and Practices, Operational Guidance for World Bank Group Staff*, The World Bank, November 2008. http://siteresources.worldbank.org/EXTENERGY2/Resources/OffgridGuidelines.pdf.

[35]Douglas F. Barnes (ed.) (2007). *The Challenge of Rural Electrification: Strategies for Developing Countries*. Resources for the Future, Washington, DC.

[36]Jim Finucane and Chris Purcell, *Photovoltaics for Community Service Facilities: Guidance for Sustainability, Africa Renewable Energy Access Program*, The World Bank, 2010. The English and French versions of the report can be downloaded from http://documents.worldbank.org/curated/en/837791468332067596/Photovoltaics-for-community-service-facilities-guidance-for-sustainability.

taking advantage of the confluence of these advances and point to a promising future. These include Mobisol, M-Kopa, Off-Grid Electric, Azuri, Sunny Money, Solar Kiosk, to name but a few.

Nonetheless, enormous challenges remain if the 500 million unelectrified rural population in Africa today (plus population growth!) is to gain access to electricity services by 2030. The challenges include lack of access to financing, sparse population (ironically sparse population makes SHS economically less costly than extending the grid but also increases the cost of serving such dispersed and poor populations), lack of infrastructure including skilled labor, unsupportive policy environment, limited access to information, and high transaction costs in doing business.

Lighting Africa, discussed below in this chapter, epitomizes a program that takes advantage of the confluence of these factors and addresses and advances the major challenges in Africa. It is contributing to a major private sector–led scale-up in solar off-grid electrification.

4.4.3   Solar PV in Africa

Increasing electricity access in Africa has always been a priority for Africans and for the development community. The share of population without access has been declining in percentage terms but has been unable to keep up with population growth. In 1992, 42% of the urban population and 92% of the rural population in Sub-Saharan Africa did not have access to electricity. In 2014, 22 years later, the share of urban and rural population without access to electricity had indeed declined to 30% and 82%, respectively. However, the number of people without electricity access increased from 225 to 294 million in urban areas and from 356 to 502 million in rural areas. By 2030, despite population growth, rural population is expected to decline to about 550 million due to urbanization.[37]

These are discouraging statistics. The rural population spent $14 billion in 2014 to purchase kerosene, candles, and flashlight

---

[37]Cities Alliance, The Urban Transition in Sub-Saharan Africa Implications for Economic Growth and Poverty Reduction. https://www.citiesalliance.org/sites/citiesalliance.org/files/CA_Docs/resources/paper-pres/ssa/eng/ssa_english_full.pdf

batteries for lighting of poor quality and often dangerous.[38] In addition, they spent millions more for charging mobile phones and buying disposable batteries for low-power appliances. The off-grid population is estimated to spend an additional $2.4 billion annual for mobile phone charging. The challenge, therefore, was how to facilitate their switch from such inferior sources to solar PV electricity.

Population location	Total population and estimates (millions)			Without access to electricity (millions)			
				Percent		Number	
	1992	2014	2030	1992	2014	1992	2014
Urban	151	363	579	41.8%	30.2%	63	110
Rural	389	612	552	91.5%	82.0%	356	502
Total	540	975	1131				

Source: World Bank, Data Bank, http://databank.worldbank.org/data/reports. aspx?source=world-development-indicators# and https://www.citiesalliance.org/ sites/citiesalliance.org/files/CA_Docs/resources/paper-pres/ssa/eng/chap_2.pdf.

Kenya represented the earliest commercial SHS market in Africa. By 2000, there were an estimated 150,000–200,000 SHS with an installed capacity of 3.5 MW in Kenya.[39] Mark Hankins, whose knowledge of the Kenyan and African solar sector is phenomenal, estimated that sales had averaged 20,000 systems annually and that the industry was worth $6 million in 2000. Hundreds of businesses were involved in solar PV importing, manufacturing, retailing and operating battery charging stations. With near-absence of electricity access in rural areas (the utility, Kenya Power and Light had less than 70,000 customers in 2000), the majority of customers were the relatively wealthy, who could afford to pay cash. One early entrepreneur was Harold Burris, who founded Solar Shamba in the 1980s. He was able to get 12 V DC appliances manufactured locally, got the local battery manufacturer to improve the design of the car battery,

---

[38]Bloomberg New Energy and Lighting Africa, Off-grid Solar Market Trends Report 2016. February 2016. https://www.lightingafrica.org/wp-content/ uploads/2016/12/OffGridSolarTrendsReport2016.pdf
[39]Mark Hankins, Kenya PV Experience, November 2001. Cited in *Sustainable Energy for All: Innovation, Technology and Pro-Poor Green Transformations* (David Ockwell and Rob Byrne), Routledge.

and trained a large number of technicians to sell and install SHS. Other companies soon got into the solar business.

In the latter half of the 1980s, the market matured and marketing and sales became more sophisticated and the range of locally made products increased. Costs, high by today's standards, were dropping. In 1985, a six-light, 40 Wp SHS was retailing for $1,000 ($25/W). In the 1990s, larger companies entered the market, responding to increased demands for systems for operating not just lights and radio, but also TVs. Smaller systems were being sold at more affordable prices; a 12 Wp SHS was being retailed at $200 in 2000 ($17/Wp). Larger markets led to sourcing complete kits, mainly from China, and the supply chains became more streamlined and mature. Consumer financing became available, though interest rates were high at about 40% and repayment periods only 1–2 years. Traditional lease-to-own companies began offering financing for SHS by 2001. The World Bank's Energy Sector Management Assistance Program (ESMAP) conducted a successful pilot financing scheme in the late 1990s, but as the Kenyan banking sector had several setbacks, the pilot project never mainstreamed. Although the IFC's PV Market Transformation Initiative offered financing for SHS companies, there were few takers as the requirements were too cumbersome and stringent.

The government of South Africa partnered with several companies to offer SHS to consumers who could not be reached by the grid as part of its Integrated National Electrification Program to achieve universal access to electricity by 2025.[40] In 1998, the South African utility Eskom and Shell Solar launched a joint venture SHS project in Eastern Cape. Currently, the partners are KwaZulu Energy Services (KES), Nuon RAPS Utility (Pty) Ltd, Solar Vision (Pty) Ltd, Ilitha Cooperative, Summer Sun Trading (Pty) Ltd, and Shine the Way cc.[41] It has provided electricity services to 50,000 households using a fee-for-service model where the monthly fee covers maintenance costs and

---

[40]http://energy-access.gnesd.org/cases/22-south-african-electrification-programme.html.

[41]GNESD, Off-grid Solar Home System Programme South Africa, http://energy-access.gnesd.org/cases/29-1-sarah-best-sustainable-development-advisors-for-the-international-institute-for-environment-and-development-2011-remote-access-expanding-energy-provision-in-rural-argentina-through-public-private-partnerships-and-renewable-energy-a-case-study-of-the-per.html.

battery replacement. The government subsidized 80% of the capital cost and offers an operation and maintenance cost subsidy.

The Global Environment Facility in partnership with UNDP was an early funder of solar PV projects in Africa dating from 1991. They provided $43 million in financing for 11 PV projects in Africa.[42] The Uganda Pilot Photovoltaic Project for Rural Electrification project tried to develop a consumer financing model by offering guarantees to Uganda Women's Finance Trust and the Centenary Rural Development Bank. However, few loans were disbursed and interest rates were high (38%). In 2001, the project attempted to use a village bank model to lend directly to consumers at lower interest rates (18%) and 1- to 2-year loan term. This project was relatively successful with 510 SHS financed—quality was good, warranties were honored, and debt recovery was 80–90%. A private company adopted this model by acquiring additional funding. A project in Zimbabwe installed more than 9,000 SHS through a subsidization scheme that offered consumer loans. Bulk procurement and exemption from duties and taxes reduced consumer prices by more than 15%. However, it distorted the domestic market and disadvantaged companies that were not participating in the project.

In several African countries, the World Bank tried different business/implementation models, such as SHS concessions in Ghana. A model area-wise electrification approach, termed Sustainable Solar Market Packages (SSMP), was attempted in Zambia and Tanzania. The SSMP offered a contract for supply, installation, and maintenance of public sector facilities for health centers, schools, government offices, etc., with an obligation to sell SHS on a more commercial basis to households. The principle behind this concept was that the sales and service of a large number of PV systems for public facilities would serve as the anchor business permitting the sales and service of SHS to be undertaken as an ancillary business. The concept was not as successful as expected. The main problem was that companies that had the financial and technical capability to design, supply, and install large numbers of PV systems (e.g., contract value about $1 million), had neither the interest nor the capability to sell and

---

[42]Martin Krause and Sara Nordstrom (May 2004). *Solar Photovoltaics in Africa: Experiences with Financing and Delivery Models*, UNDP and GEF.

service SHS on a retail basis. A World Bank review of the SSMP in Tanzania noted: "The model has not yet yielded the expected results because the companies familiar with the competitive procurements did not have the necessary corporate skill set to develop local private PV markets. An independent evaluation of the SSMP model found that not only had it failed to deliver on its promise in Tanzania, but it had also encountered the same difficulty in other countries (Zambia) where it had been tried. A demand for quality solar products exists in Tanzania and a new model for delivery of renewable energy products that minimize upfront investment costs will be supported under the proposed Project."[43]

In Uganda and Mozambique, the World Bank financed many PV systems for schools and clinics without an obligation to sell and service SHS. Later in 2015–2017, a modified version of the SSMP was undertaken under the Myanmar National Electrification Project. In this project, contractors were selected using internationally competitive bidding, and they were required to supply and install PV systems for public facilities and for households. Households paid about 10% of the cost of SHS, while the Department of Rural Development financed the costs of the public systems and the balance costs of the SHS. Maintenance services would be provided by a separate contract issued to local companies. The first procurement resulted in some of the lowest installed costs for off-grid PV systems at about $3/Wp, inclusive of lithium ion batteries and extended warranties.

4.4.4   Lighting Africa

Clean Energy Investment Framework

An important catalyst to the World Bank's support for solar in Africa was the G8 Summit held in Gleneagles, Scotland, in 2005. In July 2005, the G8 (Group of Eight Industrialized Nations) issued the Gleneagles Communiqué on Climate Change, Clean Energy and Sustainable Development. The communique recognized multiple challenges of climate change, expected growth in energy demand

---

[43]The World Bank, *Rural Electrification Expansion Program—Program for Results*, Project Appraisal Document, May 31, 2016, Page 6. http://documents.worldbank. org/curated/en/104231468184784027/pdf/103827-PAD-P153781-PUBLIC-IDA-R2016-0138-1-Box396270B.pdf.

and the consequent significant increase in greenhouse gas (GHG) emissions, the need to supply reliable and affordable energy sources, the need to concurrently reduce local pollution from energy production, and the need to increase access to modern energy. The G8 resolved to take further action to promote innovation, energy efficiency, policy, and regulations and accelerate deployment of cleaner technologies. It agreed to work to enhance access to private investment and technology transfer for clean energy deployment to developing countries. It agreed to increase the awareness of climate change and to improve access to information to better use energy and reduce emissions. It specifically requested the World Bank Group (WBG) to "take a leadership role in creating an [sic] new framework for clean energy and development, including investment and financing."

In response, the World Bank Group issued an investment framework for clean energy for development and later in 2007, an Action Plan of which one pillar was support for the energy sector, with an emphasis on the Sub-Saharan Africa energy scale-up. The Africa energy scale-up action plan intended to increase the number of households with access to modern energy to 35% by 2015 and 47% by 2030, from its low level of 25% in 2006. Soon the management recognized that achieving electricity access for 47% of the sub-Saharan African population by 2030 meant that over 500 million Africans would still have no access to electricity. That was unacceptable. It was then that Vijay Iyer, who was the sector manager for energy for the African region at the World Bank, asked me for ideas as to how the World Bank could address the electricity needs of the hundreds of millions left behind. I was at that time leading Bank-wide renewable energy work at the World Bank's Energy and Infrastructure Department. This was indeed a prescient request. Later in 2016, the UN Summit on Sustainable Development would get UN member nations to commit toward achieving universal access to electricity by 2030.

Lighting Africa was born at this moment and aimed at bringing modern lighting and other basic electricity services to the 53% of households that would not be served by the grid in 2030. Coincidentally, Russell Sturm and his team at the IFC had begun a GEF-funded project to bring off-grid solar-powered LED lighting services to Kenya and Ghana using commercial channels.

4.4.5 The Lighting Africa Program

The World Bank and IFC together proposed the Lighting Africa Program to leapfrog traditional, grid-connected electricity models in Africa. It took advantage of several disruptive technologies—solar PV, LED lighting, mobile communications, and, subsequently, mobile pay—to bring essential services of modern lighting, communications, and other services to those who could not benefit from grid electricity.

Lighting Africa is catalytic. It was designed to mobilize the private sector to reach 250 million "energy-poor" customers by 2030 with low-cost, reliable, affordable lighting services with an emphasis on renewable energy solutions. Lighting Africa would enable the energy-poor in Africa, who spend US$ 17 billion annually for purchased energy, to obtain far superior energy services using the same financial resources.

At the beginning, given the challenge of financing in rural Africa, Lighting Africa intended to focus on small LED lighting and perhaps mobile phone charging devices that were low cost, but good quality. This would allow consumers to pay cash or pay in a few installments. Later, as the market matured, the access to financing improved, including the use of mobile payments. Therefore, larger products were included within the Lighting Africa umbrella. Today, PV systems up to 100 Wp are supported, with multiple lights, mobile charging as well as other appliances such as color TVs.

Lighting Africa struck a responsive chord among the World Bank management as well as donors and industry, and as a result, getting their commitment and funding was relatively painless. Perhaps it was the simplicity of the concept, or the singular focus on lighting as an extremely important element for improving lives, or perhaps the partnership with the private sector.

Soon, the relevance of Lighting Africa to a global audience and marketplace became evident and Lighting Africa was expanded to Lighting Global. Lighting Global is now the umbrella program under which the regional programs, Lighting Africa, Lighting Asia, and Lighting Pacific, function. The industry is playing a leading role now, under the Lighting Global-supported Global Off Grid Lighting Association (GOGLA).[44] Eventually, it is expected that GOGLA will

---

[44]For more on GOGLA, visit http://www.gogla.org/.

run the Lighting Global Program. Lighting Global works with GOGLA, manufacturers, distributors, and other development partners to support the modern off-grid energy market.

4.4.6   Elements of Lighting Africa Program

We consulted with more than 100 private companies, NGOs, and other stakeholders during the Lighting Africa design phase and identified the areas where World Bank Group interventions would help to accelerate the market scale-up. Lighting Africa is responding to these requests by providing information to the lighting industry on market size, customer needs and preferences, and available distribution networks; reducing high market entry and transaction costs; addressing policy and regulatory barriers; aggregating market demand; improving access to finance; and strengthening institutional capacities.

Market Intelligence. Lighting Africa conducts market research and reports on the findings to provide data to manufacturers, distributors and retailers. It has issued 34 Market Insight reports to help companies gain access to new markets or interest new investors in these markets. Lighting Global regularly issues Off-Grid Solar Market Trends Reports, the most recent issued in 2016.[45]

At the inception of Lighting Africa, to further assess industry interest and to encourage innovation, we organized a Lighting Africa Development Marketplace (DM) competition to seek innovative solutions in off-grid lighting products and services. The competition received over 400 proposals from 54 countries, including 38 African countries, from which 20 winners received seed capital of up to $200,000 each. The winners were selected at the Lighting Africa 2008 Global Business Conference held in Accra, Ghana. The conference also provided the basic information on off-grid lighting, and a sense of the scale and needs of the market, and what is required to build business partnerships. Since then, international conferences have been held regularly

---

[45]The first report was issued in 2010 and the second in 2012. The latest report was authored by Bloomberg New Energy Finance and Lighting Global, Off-grid Solar Market Trends Report 2016, https://www.lightingafrica.org/wp-content/uploads/2016/12/OffGridSolarTrendsReport2016.pdf (February 2016).

in Nairobi, Kenya; Dakar, Senegal; and Dubai, UAE. The attendance has grown exponentially, and the quality, innovation, and scale of the products exhibited have vastly expanded. GOGLA and Lighting Global will co-host a 2018 Global Off-Grid Solar Forum and Expo in January 2018 in Hong Kong. The move to East Asia is a recognition of the crucial role played by the Asian industry in development of off-grid lighting products.

Quality Assurance. A cornerstone of Lighting Africa is its work in quality assurance. Surprisingly, at the inception, there were two schools of thought. Those at the IFC with a purer market orientation preferred a laissez-faire approach, where the market place would select winners. Based on my experiences with PV-GAP, I argued for an approach that included standards setting, testing, and enforcement. I argued that a laissez-faire approach is very risky and costly for rural African consumers and that the risk of market spoilage could set back the whole industry by many years. Fortunately, my view held. Initially we felt that simple quality standards that could be verified in developing country laboratories with limited capability would suffice. However, we agreed that greater sophistication and rigor in standards and testing would provide more credible and comparable results.

Lighting Africa developed a series of Quality Standards and testing methods to protect consumers from poor-quality products and prevent the eroding of consumer confidence in off-grid products. These standards and test methods were issued subsequently as an IEC standard 62257-9-5. Quality assurance services are now offered globally through Lighting Global as the applicability and interest in such products is now global. Today there are five testing laboratories accepted by Lighting Global for quality verification: Schatz Energy Research Center in Arcata, California, USA; Shenzhen Academy of Metrology and Quality Inspection, Shenzhen, China; Fraunhofer Institute for Solar Energy Systems, Freiburg, Germany; The Lighting Laboratory, Institute for Nuclear Science & Technology, Nairobi, Kenya; and Solar Lighting Laboratory, The Energy and Resources Institute (TERI), New Delhi, India.[46]

---

[46]For more details, see https://www.lightingglobal.org/quality-assurance-program/test-laboratory-network/.

As Lighting Africa (and later Lighting Global) has no regulatory or enforcement mandate or capability, the Lighting Africa Quality Assurance was designed to verify performance claims by manufacturers—a means for verifying truth in advertising.

Today the Lighting Global quality assurance framework includes test methods, quality standards, and a product testing and verification program. It covers markets in Asia and Africa under the Lighting Global Program. Since its launch in 2009, by June 2017, 101 products from 57 manufacturers have been quality-verified as meeting the Lighting Global Quality Standards.[47]

Access to Finance. Access to finance along the supply chain is essential to the growth of the off-grid lighting and SHS market. Lighting Africa facilitates and leverages financial products to help provide the capital that is needed by distributors and consumers. Several micro-finance institutions (four in Kenya, five in Ethiopia, two in Nigeria), and KIVA, the crowd-funding platform, are providing consumers micro-loans for quality-verified off-grid lighting and energy products. The World Bank energy access projects in eight countries, Burkina Faso, Mali, Liberia, DRC, Uganda, Ethiopia, Tanzania, and Rwanda, have integrated Lighting Africa activities. The Development Bank of Ethiopia with World Bank funds established a foreign exchange credit facility to support the import of Lighting Global–qualified products. As a result, 800,000 quality-verified products were imported into Ethiopia to benefit 1 million Ethiopians.

Consumer Education. Lighting Africa provides unbiased information to users on the benefits of good-quality off-grid lighting and energy products. It does so by generating and disseminating consumer education materials. It also cooperates with manufacturers and distributors in its outreach efforts.

Business Development Support. Lighting Africa provides advice on good business practices and risk management. It helps develop and inform the industry of new business models that are effective in reaching the poorest consumers in Africa so they can obtain good-quality affordable products.

Lighting Africa also works with governments toward *removing policy and regulatory barriers* to increase access to clean energy; to foster a competitive market for off-grid energy

---

[47]See product list at https://www.lightingglobal.org/products/.

products; and to integrate modern off-grid services into their rural electrification programs. The governments of Ethiopia, Kenya, and Tanzania and the Economic Community of West African States (ECOWAS) have adopted or are in the process of adopting national standards for off-grid solar products that are harmonized with Lighting Global Quality Standards.

Lighting Global is operating in Burkina Faso, DRC, Ethiopia, Ghana, Kenya, Liberia, Mali, Nigeria, Senegal, South Sudan, Tanzania, and Uganda in Africa. In the Asia-Pacific region, it is functioning in Afghanistan, Bangladesh, India, Myanmar, Pakistan, and Papua New Guinea.

The results to date are encouraging: As of June 2016, an estimated 131 million people globally have benefitted from Lighting Global Quality–verified solar products.[48] Over 26 million quality-verified lighting products have been sold globally in 75 countries since 2008. There are 121 quality-verified products on the market. The off-grid solar industry had $276 million invested in 2015, a 15-fold increase since 2012. It is expected there will be a $3.1 billion market opportunity for the off-grid solar industry by 2020, reaching 99 million households. The United Nations Framework Convention on Climate Change (UNFCCC) is requiring solar lighting products to meet the Lighting Global Quality Standards to qualify for carbon financing (CDM).

4.4.7 Lessons Learned

Many general lessons have emerged from the World Bank Group's experiences—some seem self-evident, but are too frequently overlooked. Among them are the following[49]:

- Best practices can be rarely implemented in their entirety. Especially in countries with weak institutions and infrastructure, some crucial elements may be not practically implementable. However, this does not mean such best

---

[48]Lighting Africa, Our Impact, https://www.lightingafrica.org/about/our-impact/.

[49]Excerpted from The World Bank, *Beyond Bonn: World Bank Group Progress on Renewable Energy and Energy Efficiency in Fiscal 2005 –2009*, The International Bank for Reconstruction and Development/The World Bank Group, December 2009, and other sources.

practice requirements can be ignored. Some work-around ways must be found to achieve best practice elements.

- Money alone will not bring change. Efforts should also focus on capacity building and good governance. That said, without financial resources, scaling up of solar and renewable energy will not happen. Moreover, synchronization between investments and capacity building is needed. If timely investments do not occur, the capacities already built will quickly dissipate.

- Governments must be market enablers, create a favorable business and regulatory environment, and adopt transparent decision-making mechanisms. Energy sector reform is sine qua non if investments are to pay off and be scalable and sustainable, and if the benefits are to flow to those most in need.

- It is necessary to engage with private sector interests, including project developers, technology suppliers, and entrepreneurs, to obtain their perspective on market needs. What works in one country at one time may not be what is needed elsewhere or at another time.

- Given the immense challenges at hand, increased coordination among multilateral and bilateral agencies, governments, nongovernmental organizations (NGOs), the private sector, and community groups is imperative to avoid duplication of projects and programs. This is far easier said than done, as each organization has their own agenda and schedule and a reluctance to cooperate with others.

- Ensuring financial and economic viability is of paramount importance, or projects will not be sustainable in the long term. Donors and governments should be cautious about introducing, heavily subsidized programs. These efforts may provide a small short-term boost but may not support the long-term development of nascent commercial markets, and they often destroy the capacity of the developing country entrepreneurs to deliver alternative energy products and services profitably and sustainably.

- Capital investments must be linked closely to committing resources and building capacities to maintain facilities and provide reliable and useful service in the long term. Too often, the provision of maintenance and repair services is overlooked or the challenge is underestimated.

- Innovation is needed. Given the enormity of the challenge and the huge resources needed, it is necessary to find low-cost ways of sustainably providing these energy services and more rapid ways of building the infrastructure and human capacity needed to successfully introduce the technology.

- Keep it simple and ambitions within bounds. Avoid overly complex project designs with too many objectives and overly ambitious goals, and requiring many actors to cooperate and coordinate.

- Finally, and most important, sustained political commitment to action by all partners is needed. Good intentions alone are insufficient.

4.4.8   The Future

Historically, World Bank Group investment in renewable energy in developing countries was a relatively small share of global investments in that sector. From about 1% share of global renewable energy financing in developing countries in 2005, it has trended downwards and it is about 0.3% of renewable energy investments in developing countries in 2016.[50] In 2004, World Bank Group renewable energy lending to developing countries was US$ 1 billion, while total global financing was about US$ 10 billion. In 2016, World Bank Group lending was US$ 1.7 billion compared with the global lending of US$ 117 billion for renewable energy in developing countries.

In the early days, the World Bank's role as a catalyst and in building confidence in renewable energy was far greater than the magnitude in financing in mere dollar terms. In the future, the World Bank Group needs to strategically invest so that its funds are leveraged many times over; this is not to diminish the importance of World Bank Group investment financing. This will require a change in how funds are deployed. World Bank support could help improve the policy and regulatory

---

[50]Sources: Frankfurt School-UNEP Centre/BNEF (2017). *Global Trends in Renewable Energy Investment 2017*, http://www.fs-unep-centre.org (Frankfurt am Main); World Bank Group data.

environment and investment climate. It can help mitigate investment risks by using World Bank Group guarantee instruments, co-finance investments with development partners and the private sector, and improve the quality of projects through advisory services. It can help improve the institutional capability in developing countries' government agencies, regulators, the banking sector, businesses, and electric utilities. It can offer technical assistance to improve planning capabilities, including integration of variable renewable energy technologies such as wind and solar, to address quality concerns and to support technology transfer. The Lighting Africa program described previously is an example of how several of these instruments are applied to leverage investments in renewable energy.

4.4.9 Paris Climate Agreement (2015)

The United Nations Climate Change Conference, COP 21, was held in Paris, France, in December 2015. The conference negotiated the Paris Agreement to mobilize the global community to respond to the threat of climate change by keeping a global temperature rise this century well below 2°C above pre-industrial levels and to pursue efforts to limit the temperature increase even further to 1.5°C. The agreement also helps countries to deal with climate change impacts.

The Paris Agreement is a voluntary agreement to get participating countries to commit to and achieve "nationally determined contributions" (NDCs) and report regularly on their emissions and on their implementation efforts. To support developing countries to meet their commitment, increased access to financing, technology transfer, and capacity building would be facilitated.

On October 5, 2016, the threshold for entry into force of the Paris Agreement was achieved and the Agreement entered into force on November 4, 2016. Thirty-nine Sub-Saharan African countries submitted commitments that proposed grid and off-grid solar solutions, among others, to meet their greenhouse gas emissions reductions targets.[51] These are shown in Fig. 4.2.

---

[51]For details see http://spappssecext.worldbank.org/sites/indc/Pages/INDCHome.aspx.

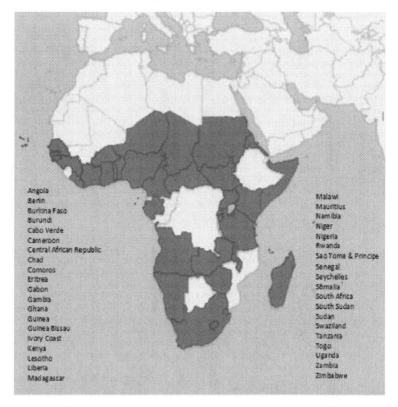

Figure 4.2 Sub-Saharan African countries committing to meet NDC targets using solar.

4.4.10   Climate Change Action Plan 2016–2020

The IFC estimated that the Paris Climate Agreement will require more than $23 trillion in renewable energy and energy efficiency investments in 21 emerging markets representing 62% of the world's population and 48% of global GHG emissions.[52] This represents an average financing requirement of US$ 1.5 trillion a year until 2030. If the World Bank Group is to

[52]The IFC estimates that key sectors in these countries have an initial investment opportunity of nearly $23 trillion from 2016 to 2030. This figure is likely an underestimate, as there are data gaps for important sectors such as climate-smart agriculture. See IFC (2016). *Climate Investment Opportunities in Emerging Markets: An IFC Analysis*, https://www.ifc.org/wps/wcm/connect/51183b2d-c82e-443e-bb9b-68d9572dd48d/3503-IFC-Climate_Investment_Opportunity-Report-Dec-FINAL.pdf?MOD=AJPERES.

retain its share of financing through 2030, then the average financing the World Bank Group would need to provide must rise to US$ 46 billion a year, or 27 times the financial commitment made in 2016. Clearly, this is unlikely. In 2016, the World Bank Group committed about $64 billion in loans, grants, equity investments, and guarantees to its members and private businesses in *all* sectors, of which 21% (US$ 13 billion) was climate related.[53] Even with far greater capital replenishments, due to many other priorities, the share of World Bank Group lending for climate-related investments may not reach the levels seen in 2016, or from a global perspective, the share achieved in 2005.

To help the countries meet their NDC commitments, the World Bank Group issued the Climate Change Action Plan 2016–2020.[54] The Action Plan commits the World Bank Group to increase its climate-related investments in its portfolio from 21% to 28% by 2020. It expects to finance and *leverage* financing of up to $29 billion per year by 2020 for climate-related investments. Given the enormous financial requirements, this anticipated level of commitment will see a diminishing role for the World Bank in helping developing countries meet their Paris Climate Agreement commitments. Therefore, the World Bank Group needs to be smarter and more strategic in how it utilizes its own resources and, yes, commits more resources, to leverage an order of magnitude higher level of financing from other sources to help developing countries meet their NDC targets.

World Bank Group activities in their Climate Change Action Plan fall within four priorities:

- **Support transformational policies and institutions.** The World Bank Group will support its partner countries to achieve their NDCs by assisting countries in establishing the right policies, developing investment plans, and ensuring climate considerations are integral to their planning and budgeting processes.
- **Leverage resources.** The Group will increase its own investments in clean energy and leverage them with private sector resources. It will help create the enabling

---

[53]World Bank Group Annual Report 2016, file:///C:/Users/Anil%20Cabraal/ Documents/Downloads/9781464808524.pdf.

[54] World Bank Group, Climate Change Action Plan 2016-2020, https://openknowledge. worldbank.org/bitstream/handle/10986/24451/K8860.pdf?sequence=2.

environment to support climate-friendly investments. It will also ensure adequate funds are committed to climate-friendly investments and help countries mobilize additional climate financing.

- **Scale up climate action**. The Group actions will concentrate on six focus areas, including renewable energy, and scale up its advisory and investment services.
- **Align internal processes and work with others**. The Group's internal procedures and incentives will be aligned to support climate-friendly investments, including risk screening for climate impacts, reporting on the impact of operations on GHG emissions, providing easier access to data related to climate, among others.

The World Bank Group has committed to specific goals for renewable energy over this 5-year period. It committed to directly financing 20 GW in renewable energy generation over 5 years, which is double the current World Bank Group–assisted capacity additions.[55] The World Bank Group will support strengthening transmission grids to catalyze an additional 10 GW of variable renewable energy with other sources of financing. It will help ensure that all energy investments are adapted to climate change. The IFC and the Multilateral Investment Guarantee Agency (MIGA) will support grid-connected clean energy by focusing on large hydropower, wind, and solar projects.

4.4.11   IFC Scaling Solar

The rapid reductions in costs and competitiveness of grid-tied solar PV systems has resulted in intense interest of the World Bank Group client countries in large-scale solar PV projects. To respond to these demands, the IFC launched Scaling Solar to provide World Bank Group services aimed at creating viable markets

---

[55]These commitments are conditional on sustained World Bank Group lending volumes, access to adequate concessional financing, and client demand. While this target may be relatively large for the World Bank Group compared with historical achievements, from a global perspective it is small today, as global renewable energy investments reached nearly 140 GW in 2016 with about half in developing countries.

for grid-tied large solar systems, primarily in Africa.[56] The "one stop shop" program aims to make privately funded grid-connected solar projects operational within 2 years and at competitive tariffs. When implemented across multiple countries, the program will create a new global market for solar investment. Scaling Solar support targets specific challenges in scaling up grid-tied solar investments in developing countries. Specific constraints addressed are as follows: limited capability of governments to manage, structure, and negotiate private power concessions; lack of scale, which deters investors; limited or no competition, which increases risk and costs for both investors and governments; high transaction costs of individually negotiated contracts; and high-risk perceptions.

To overcome these constraints, Scaling Solar offers a package of services, including advice to assess the right size and location for solar PV power plants; simple and rapid tendering to ensure strong participation and competition from committed industry players; standardized templates of bankable project documents that can eliminate negotiation and speed up financing; competitive financing and insurance; and risk management and credit enhancement products.

Scaling Solar has engagements in Zambia (successfully tendered 2 × 50 MW solar projects and a follow-on agreement signed for another 500 MW). The first received a bid for 6.02 U.S. cents per kWh flat rate for 20 years). Other engagements are in Senegal to develop 200 MW; Madagascar to develop 30–40 MW; and 500 MW in Ethiopia.[57]

4.4.12   World Bank Off-grid Solar Projects

World Bank has implemented and is implementing several solar PV electrification projects in Africa, including support for solar home systems as well as mini-grids. All followed a similar model of offering concessional financing and/or subsidy and supporting market development and capacity building. The most recent is

---

[56]*Scaling Solar, Unlocking Private Investment in Large-Scale Solar Power*, An IFC Program. http://www.ifc.org/wps/wcm/connect/811dab004b378c4f8675fe4149c6fa94/ ScalingSolar2016.pdf?MOD=AJPERES
[57]Scaling   Solar,   Active   Engagements,   http://www.scalingsolar.org/active-engagements/#tab-id-4.

in Rwanda, where a proposed project will establish a $50 million credit line for financing SHS and mini-grids. In Ethiopia, an off-grid renewable energy program will provide carbon financing obtained under the Clean Development Mechanism for solar lamps and SHS. At least 50,000 households are targeted for the first year, and the government envisions that 3,000,000 solar lamps and SHS will be supported by the end of the program. In Liberia, the Renewable Energy Access Project anticipates providing off-grid electricity access to 100,000 people. The 2016 Senegal Rural Electrification Program anticipates that nearly 600 villages will receive electricity services from solar PV mini-grids and SHS.

In conclusion, solar photovoltaic technological innovations, cost reductions, and new business models have increased confidence among African consumers and investors in solar photovoltaics as a means for providing electricity services to millions of consumers. This has revitalized the off-grid electricity service industry and made possible the goal of bringing universal electricity access to Africa by 2030, or even sooner. Lighting Africa has played a crucial role in making this possible.

4.5   The Africa Clean Energy Corridor

As an intergovernmental organization, one of IRENA's strengths is its convening power. Several times per year, delegates from close to 100 nations come together in Abu Dhabi to discuss various pathways progressing their renewable energy agendas. This enables debate and coordinated action on a national, regional, and global level.

At one of the IRENA gatherings in Abu Dhabi in 2012, Gosaye Mengistie Abayneh, then director of Energy Studies & Development at the Ethiopian Ministry of Water and Energy, together with Steve Sawyer, secretary general of the Global Wind Energy Council, first formulated the idea to create a pan-African electricity network stretching from Cairo to Cape Town. They had analyzed how in Ethiopia a natural complementarity between seasons of wind and hydropower exists. Gauri Singh, in charge of the Country Support and Partnerships division at IRENA, continued the analysis for a corridor stretching from Ethiopia and Kenya down to Mozambique and South Africa. In that process, Abayneh and Singh realized that apart from seasonal variation, there was also variation in demand and loads that could not be appropriately matched with the present generation and connection capacity. For a more optimized electricity system, a conceptual framework was formulated with stronger and smarter grid interconnectivity, also allowing for more renewable energy capacity. This was the birth of the Africa Clean Energy Corridor.

The tremendous economic growth of many African countries and the additional energy requirements, combined with the increasing competitiveness of renewable energy technologies, requires new thinking. In the early 2000s, when oil prices were very low, many African countries had installed diesel generators for their electricity needs. Continuously rising oil prices had, however, made this a very expensive solution and many governments were looking for alternatives. They realized that clean renewable power sources such as hydropower, geothermal power, biopower, wind power, and, of course, solar power could help to reduce dependence on fuel imports, fossil fuel consumption, carbon emissions, and electricity costs. This could also help to expand access to electricity and create new jobs in the rapidly growing economies of Eastern and Southern Africa. In addition, there was a growing recognition of the benefits of integrating electricity networks beyond national borders.

There are many synergies to be obtained in linking regional energy markets, specifically electricity markets. Many countries are connected and organized through power pools. In Africa, for example, there are several regional power pools, with functioning wholesale markets, including the Eastern Africa Power Pool (EAPP) and Southern Africa Power Pool (SAPP), with the aim to facilitate cross-border trade of electricity. Figure 4.3 shows the extent of Africa's power pools. Expanding renewable capacity in such markets can benefit from cross-border sales, leading to substantial savings and a higher overall level of supply security. It is projected that the overall demand in Eastern and Southern Africa would grow an average of 5.8% per annum from 2013 to 2038, more than quadrupling from 243 billion to 983 billion kWh in those 25 years; associated needs for new generating capacity are over 150 GW. The Southern Africa Power Pool projects that the peak load will grow an average of 4.2% per annum until 2025, more than doubling from roughly 41 to 94 GW over 20 years; projected needs for new generating capacity over the period are nearly 57 GW.

After its conception, the **Africa Clean Energy Corridor** initiative was first formally introduced at the third IRENA Assembly in January 2013. At the time, Frank Wouters was deputy director-general of IRENA and the Africa Clean Energy Corridor was close to his heart. Gauri Singh and her team, including Jeffrey Skeer and Safiatou Alzouma, regional programme officer Southern Africa, realized that such a grand vision requires careful thinking and a smart strategy for implementation. They knew that if they would go straight to the energy ministers, they might face opposition from the real decision makers, the professionals working at the local utilities, the regulators, power pools, etc. They knew that if this group would not support the idea and philosophy behind it, they had endless power to ignore the idea, delay its implementation, and generally frustrate anything going forward. So, IRENA devised a two-stage approach. First, they would get the real decision makers on board and organized a two-day workshop on June 22–23, 2013, in Abu Dhabi, which brought together representatives of major regional bodies, power pools, utilities, independent power producers, ministries, financial institutions, and development partners to discuss what could be done to increase the share of renewables in future generation plans. The workshop was a big success, the turnout was impressive, and everybody participated and stayed until late in the meeting room. Even on the second day, everybody turned up on time and stayed for the entire meeting; nobody left to visit one of the many grand shopping malls of Abu Dhabi! But more important, people understood the significance of this initiative, which could become the foundation of many future renewable energy initiatives, programs, and projects.

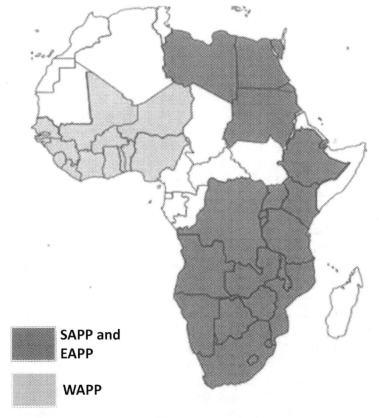

Figure 4.3 Africa's power pools.[58]

---

[58]Tijana Radojičič, IRENA, Dar Es Salaam, March 9, 2016.

After IRENA had secured the buy-in from the work floor so to speak, they could go a level higher. They managed to convince Ms. Nkosazana Clarice Dlamini-Zuma, chair of the African Union and wife of South African President Jacob Zuma, to include the Africa Clean Energy Corridor in a speech held at the United Nations. The initiative was consequently endorsed by ministers from the countries of the Eastern Africa Power Pool and the Southern Africa Power Pool at the fourth IRENA Assembly in January 2014. In March 2014, African finance ministers welcomed the idea at a gathering in Abuja, Nigeria, and at the United Nations Climate Summit of September 2014, the Corridor concept was presented to the United Nations.

So, with the political process for the initiative moving ahead expeditiously, now the fundamental issues had to be tackled.

4.5.1 The Issue at Hand

Part of the problem with regional energy planning is that often the data that lies at the core of an energy plan or strategy evolves very dynamically, which is especially the case for renewable energy. There is tremendous progress in the field of technologies and costs, and they can substantially differ over the course of even a few years. According to IRENA, electricity tariffs typically range from 6 to 17 U.S. cents per kWh in Eastern Africa and 5 to 16 cents in Southern Africa based on coal, oil, gas, and hydropower generation. The efficiency and cost for coal-fired power or hydropower has not changed dramatically over the past couple of decades, so it is easy to make a forward-looking plan. Solar and wind power, however, now only cost a fraction of what they used to cost even a few years ago and are competitive with conventional power. There are not many countries that have taken a correct forward-looking view of the cost of solar and wind power in their energy planning. A regional plan is typically built up using national plans. Considering that national plans must be approved by the government and sometimes even parliament, one can understand that a regional plan is always based on data that is several years old. For example, the Eastern Africa Power Pool's 2011 Master Plan foresees adding 24.6 GW of hydro capacity through 2038 but foresees just 7.9 GW of wind, 3.4 GW of geothermal, and 50 MW of solar capacity. The Southern Africa

Power Pool's 2009 Pool Plan includes some hydropower but no other renewable capacity additions. The fact that just a few years ago the official planning for almost an entire continent with tremendous solar potential foresaw just 50 MW of solar PV shows the weakness of such inherently slow planning processes, especially since solar PV electricity is now in most places the cheapest and quickest-to-install form of power.

The Africa Clean Energy Corridor initiative intends to help countries and Power Pools by providing tools and support to arrive at a more accurate picture for Africa's future energy system, incorporating much more renewables than before. The addition of low-cost renewables will reduce cost and price volatility and increase security of supply. The idea behind the Africa Clean Energy Corridor was to develop and implement a vision for an integrated African renewable energy future, stretching from north to south. The Eastern and Southern African Power Pools would work with national stakeholders to increase the share of cost-effective and clean renewable energies and create a pan-African market.

In the Africa Clean Energy Corridor initiative, we included several elements, ranging from planning, resource assessment, and access to finance to support on technical aspects of modern renewables.

4.5.2 Planning

Grid integration of a high share of renewable power, particularly variable wind and solar power and run-of-river hydropower, requires careful planning. Power pools and grid operators must keep the grid stable and maintain power quality to ensure a reliable power supply to their customers. To increase the share of variable renewables beyond a certain threshold, depending on the strength of the grid in place, a variety of options may be applied. Operational options include better load forecasting and coordination with generation at hydropower dams and gas-fired plants, whose output can be adjusted rapidly, to compensate for wind and solar fluctuations. Technology options include smarter grids to better regulate power flows and storage through pumped hydro or advanced batteries and stronger interconnectors to reduce the variability of power supply on a regional basis.

Coordinated planning at country and regional levels can help power pools and power producers choose the most cost-effective strategies for ensuring reliable service as renewable power shares increase.

Joint planning of generation and transmission facilities would yield a better power grid at lower cost. Currently, generation expansion plans are formulated by each country in the Eastern and Southern African Power Pools independently, and the Power Pools aggregate them. The Eastern Africa Power Pools' 2011 Master Plan found that joint regional optimization of generation and transmission plans could save US$ 7.3 billion over 25 years on top of the $25.2 billion of savings that are derived from optimization within each country separately. The Southern Africa Power Pool's draft 2009 Pool Plan found that coordinated planning could save US$ 47.5 billion over 20 years, of which a major share also would result from regional planning. Since these plans do not even include savings from cheap solar and wind power, the actual savings will be much higher. Coordination of generation and transmission planning between the Eastern and Southern Africa Power Pools could yield further operational and cost efficiencies.

But there is more to gain. Coordinated zoning of renewable and other generating facilities could provide significant economies of scale in transmission infrastructure. High-voltage transmission lines can move large amounts of electricity at a low cost per kilowatt-hour. With large amounts of power clustered in renewable energy zones based on resource potential, cost-effective high-voltage transmission lines can be built to move the electricity generated to major load centers such as mines, other industries, and rapidly growing cities.

In areas that are not connected to the main power grid, inclusion of renewable power options in mini-grids can reduce generating costs, expand access to electricity services, and create stable building blocks for extension of the main power grid in future.

4.5.3 Resource Assessment

Together with the Masdar Institute, IRENA has developed the Global Atlas for Renewable Energy (irena.masdar.ac.ae), a free online resource-assessment tool intended to help policy makers and investors appreciate the extent of the renewable energy resources at their disposal in each country or region. Many IRENA member states make their renewable energy resource data available to enrich the tool. IRENA is working with countries in Eastern and Southern Africa to help fund production of bankable data on their wind, solar, and biomass resources, as well as exploration and assessment drilling for their geothermal resources. The aim is to reduce the investment risks of developing these resources into cost-effective power projects. IRENA is, furthermore, partnering with countries and expert institutions to help identify suitable zones for concentrated development of renewable power resources, as described above, and to facilitate planning for the joint optimization of investments in generation and transmission.

4.5.4 Access to Finance

Access to finance is crucial and for many African countries not easy, mostly due to lack of skills and capacity to develop bankable projects and the general risk(s), real or perceived, associated with any investment in the region. IRENA is working with multilateral financial institutions, project developers, and electricity market regulators in Eastern and Southern Africa to develop a Project Navigator (navigator.irena.org) to help investors move from renewable resource assessments to concrete power projects. The Project Navigator provides tools for the development of bankable project proposals and assists in their application. IRENA is, for example, working on model Power Purchase Agreements (PPAs) and credit enhancement tools, including de-risking mechanisms such as partial risk guarantees and credit insurance.

4.5.5 Status and Way Forward

Many things have happened since the inception of the Africa Clean Energy Corridor. Not only has the Africa Clean Energy Corridor taken firm shape, West-Africa also initiated its own West Africa Clean Energy Corridor (WACEC), a joint initiative of IRENA with the ECOWAS Centre for Renewable Energy and Energy Efficiency

(ECREEE), the West African Power Pool (WAPP), the ECOWAS Regional Electricity Regulatory Authority (ERERA), and the 14 countries of the West Africa Power Pool region, the African Development Bank, the Sustainable Energy for All (SE4All) Africa Hub and other development partners.

The WACEC's implementation is built on the same five pillars of the clean energy corridor concept: resource assessment and zoning, national and regional planning, enabling frameworks for investments, capacity building, and public and political support.

In March 2017, the first session of the Specialized Technical Committee on Energy, Transport and Tourism of the African Union recommended the African Union member states to integrate the concept of the Clean Energy Corridors into national agendas. This included ACEC for the East and Southern Africa regions and WACEC for West Africa. They were recommended to integrate the concept into their national renewable energy and climate change agendas as well as the process of creating sustainable and low-carbon power markets, as part of the concluding declaration, which will be presented to the heads of states for endorsement.

This political momentum builds on the work that has been achieved. In 2014 and 2015, the zoning methodology was developed and high-resource-potential zones for the development of solar (CSP, PV) and wind power plants in all the EAPP and SAPP member countries were identified. Least-cost system planning test models to support long-term power system expansion plans in African countries were also developed.

IRENA worked with Swaziland and Tanzania in a little more detail to consider cost-effective zones for solar and wind for the development of a more integrated national energy and power master plan.

To support national and regional agendas for sustainable power system development, IRENA initiated a Planning Oversight Project. The aim is to assist countries and regions with power system planning, by promoting enabling governance frameworks based on global good practices.

To support project initiation, development, and financing, the Sustainable Energy Marketplace (SEM) was launched by IRENA's KPFC division, which is also relevant for Corridor projects and initiatives. Several capacity-building activities targeting the relevant national stakeholders were organized, including the Renewable Energy Training Week on Regulation, training seminars on the zoning process, capacity building on energy planning and the use of planning tools for the ACEC countries. And when the projects reach a certain level of maturity, the financial viability of several sites earmarked for project development within selected zones will be assessed using standard financial feasibility metrics.

The zoning methodology can further identify high-importance projects, which can be considered in the update of the regional master power master plans as well as the Programme for Infrastructure Development in Africa (PIDA).

Lastly, new developments on the ground will necessitate a regular review and update of the zones that were identified in the 2014–2015 zoning exercise.

In West Africa, a scoping study for the solar component of the WACEC, aiming to install 2 GW of solar by 2030 in West Africa, funded under the ongoing European Union Energy Initiative's Technical Assistance Facility, was completed.

The strength of the Africa Clean Energy Corridor initiatives lies in the unifying approach across national borders, the focus on building enabling infrastructure, and, of course, the focus on thus far underdeveloped renewable energy generation capacity. The political support and the fact that development partners are part of the process also provide a pathway for finance, a crucial element that many countries still struggle with.

4.6   Global Energy Transfer Feed-in Tariff (GET FiT)

The possibility of a new unexpected large solar PV market was initiated in Africa on December 12, 2016, as the result of a scheme developed by the German Deutsche Bank and the KfW Bank.

Sub-Saharan Africa (not including South Africa) is inhabited by 816 million people. In Sub-Saharan Africa, 590 million people have no access to electricity[59]: 90% of the population in 4 countries, 80% in 5 countries, and 50% in 16 countries have no electricity. The population without electricity mostly live in small communities or are scattered. Many countries in the developed world, international organizations, and NGOs have been trying to develop schemes for providing these people with electricity, which would improve their living standard. One scheme of providing electricity to individual households and small communities is described in this section of this book. Another scheme described in Section 6 of this book is to deploy large-scale electricity-generating systems to enhance the existing central power stations. KfW developed and is introducing a third scheme to expand the utilization of the feed-in tariff (FiT) system to possibly in all Sub-Saharan African countries.

The FiT system was introduced in Germany in 2000 and refined in 2004 to encourage investors to support the establishment of electricity-generating systems powered by RE. This scheme became so successful that a large number of other countries also adopted it. A few countries in Africa introduced their own FiT systems, but they were mostly not successful.

The following text describes what an FiT system is and the brief history of how and why the German FiT was established:

The terrestrial PV industry in 1973 achieved a dependable solar cell and module manufacturing technology to produce reliable and long-lasting PV modules, the performance of which could be guaranteed for at least 20 years, and many markets for PV products were developed. What was not achieved was a sustainable huge market which—as it was predicted with

---

[59]http://data.worldbank.org/indicator/EG.ELC.ACCS.ZS.

the technology existing at that time—would support the mass production of PV modules to achieve very low prices. The U.S., Japan, and EU governments came up with programs to support solar roofs for thousands, hundred thousands, and millions of homes, but investors did not consider the government programs sustainable and did not invest money in plants to mass produce PV modules.

As already mentioned, in 2000 the FiT system was first developed in Germany. A very simple FiT system was formulated so that neither the government nor the utilities should incur any expenses. The government would not provide any financial support. It simply declared how much the electric utility would pay per kWh guaranteed for 20 years for the electricity produced by the RE (e.g., photovoltaic) system fed into its grid. The FiT was calculated so that the investor should recover in 20 years the money invested to establish the RE system and also earn a reasonable profit. This would not cost the utility anything, as it would add the surplus paid for the RE electricity to the base rate they charged to their customers. In this way, the utility would not lose money and the increase in tariff would be minuscule. The German population accepted this system.

The feed-in tariff was reduced periodically with the idea that the increased demand would result in the mass production of the PV modules leading to price decrease. The system was successful, and PV module prices plummeted: $3.82/W in 2000 and below $0.50/W in 2017. So by 2015 the electricity produced by PV systems was equivalent or even less expensive than the electricity produced by conventional systems, especially nuclear or coal.

A few of the Sub-Saharan states, as already mentioned, attempted to establish their own Renewable Energy Feed-in Tariff System (REFiT) but were not successful. The problem was that these countries had budget constraints and were also unable to receive any investment to build RE electricity-generating systems. They did not receive investment domestically and from abroad because there is a huge difference in investing to sell electricity to the local utility in a developed country, e.g., Germany or France, and doing so in a developing country. The main issues that generally inhibit private investment in renewable energy in developing countries is the lack of creditworthy off-taker structures, significant (perceived) political risk, and a lack of financial viability of investments due to low feed-in tariffs.

The Advisory Group on Energy and Climate Change (AGECC) of the Secretary General of the United Nations had asked Deutsche Bank's experts to present new concepts for promoting renewable energy investments in developing regions. The "Global Energy Transfer Feed in Tariff" (GET FiT) scheme was developed by experts of Deutsche Bank Climate Change Advisors and presented in January 2010.

- A very important and excellent idea was that the size of the project supported by GET FiT should be in the 1–20 range or maximum 50 MW, positioning it between small home systems (SHS) and utility-scale systems. That would make the raising of capital easier.
- They believed that hydropower projects would yield assured good results and therefore they would be easier to finance. The 1–20 MW range was also good as it would include many smaller hydropower projects.

In order to make the FiT program in a developing country acceptable to investors, the GET FiT concept needed the following:

- The developing country's government and/or the country's Electricity Regulatory Authority (ERA) should support RE projects by implementing appropriate regulatory schemes for renewable energy generation.
- The financial viability of investments should be established in spite of low feed-in tariffs.
- A fair risk allocation between the public and the private sector should be established.

Deutsche Bank in cooperation with the German bank KfW developed a feasibility study for a GET FiT pilot program for Uganda. Uganda was an excellent choice among the Sub-Saharan countries to start the GET FiT program. It was an excellent selection, because Uganda is one of the countries with the lowest per capita electricity consumption in the world, with 215 kWh per capita per year (Sub-Saharan Africa's average: 552 kWh per capita, world average: 2,975 per capita); therefore it required quick help to improve the situation.

The other reason is that the country has excellent technical people. This was established in connection with a study made in 2003 for the government of Uganda (GoU) and the United Nations Development Program (UNDP) for the assessment of what would be needed to establish a PV quality testing facility in Uganda.[60] When the program was started, a private PV market already existed in Uganda serviced by a surprising number (51) of private companies. The Uganda Renewable Energy Association (UREA) members consisted of 6 consultants, 3 PV manufacturers, and 45 sellers and installers (including 3 involved in solar thermal, 1 in wind, and 1 in biomass). Solar Home Systems from 10 W up to 150 W were mainly sold to the "richest" of the poor people in rural areas. At least a total of a few thousand installations were made in the years before the study.

In parallel with the private market, the Uganda Photovoltaic Pilot Project for Rural Electrification (UPPPRE) was established in 1997 and funded by the UNDP with $1.7 million.

Uganda had a National Bureau of Standards (UNBS), established in 1989, which is a member body of the International Organization of Standardization (ISO).[61] The UNBS had excellent technical people and they were very knowledgeable about PV products and systems. By 2003, the UNBS established several PV standards. Two of them—"Code of Practice for Installation of Photovoltaic System" and "PV Battery"—were so novel that they were accepted as international standards.

Uganda's present list of standards includes many PV standards. In a report dated June 14, 2017, the UNBS indicated that it is actively enforcing those standards. It stated that it destroyed substandard goods, including the PV modules entering the country that did not meet the standards.

KfW launched the GET FiT program in Uganda in May 2013. The first task for KfW was to establish cooperation with the GoU and the Electricity Regulatory Agency (ERA). In order to ensure rapid and efficient implementation of the program, the GoU provided KfW with delegated authority in terms of implementing the program. The ERA has fully embraced the program. Its staff

---

[60]UNOPS GLO/96/109-PS 120409—Peter F. Varadi, consultancy services for a PV testing facility assessment in Uganda.
[61]https://www.iso.org/member/308866.html.

to support the KfW program put in considerable effort to ensure full compatibility with its own planning system.

As already mentioned, the GoU REFiT,[62] being too low, was not attractive to investors. As a solution, KfW established a scheme in which a GET FiT "Premium Payment" (GFITPPM) would be added to the Uganda REFiT tariff levels as published by the ERA for small-scale RE projects (between 1 and 20 MW installed capacity) promoted under the REFiT system. The GFITPPM payment would be available for 5 years on a grant basis. The governments of Norway, Germany, the United Kingdom, and the European Union provided funds for GFITPPM (Table 4.1).

Table 4.1 List of donors and their contributions to GFITPPM

Donor	Net amount committed (US$)
European Union	23,000,000
Germany	17,821,095
UK	48,900,969
Norway	17,929,046
Total	**107,651,110**

The last problem to solve was the risk for investors. The risk could be political or financial (e.g., liquidity). The GoU officially requested the World Bank to use the Partial Risk Guarantee (PRG) mechanism for short-term liquidity support and, furthermore, cover for other obligations of the government and for political risk insurance (Multilateral Investment Guarantee Agency—MIGA; see footnote 28). These guarantees would be direct to lenders for projects.

The KfW scheme to add GFITPPM and the World Bank insurance to the Uganda REFiT was adequate for lenders to invest in the projects identified by KfW.

As a result of the cooperation between the ERA and the GET FiT organization, 16 projects ranging 1–20 MW systems were selected. Figure 4.4 indicates the locations of the planned 16 projects, of which 13 are hydropower projects. As mentioned, KfW considered primarily hydropower projects for the GET FiT

---

[62]http://www.GETFiT-reports.com/2013/get-fit-uganda/who-has-rallied-behind-get-fit/.

program. One bagasse project was added because Kakira Sugar Company had already one in operation and the KfW project was its extension. Two solar PV projects were added on the GuO's request. By the end of 2016, 11 projects were either in construction or commissioned. Nine hydropower projects were in construction and two projects—a bagasse and a solar PV—were commissioned. The second PV project was started in early 2017.

Figure 4.4 GET FiT projects in Uganda.[63]

4.6.1 Hydropower Projects

The GET FiT project was initiated in 2013 and, as mentioned, 13 hydropower projects were planned, of which nine were in construction phases as of the end of 2016.

From Table 4.2, it is evident as of the end of 2016, none of the initiated hydropower projects were completed. The advantage

---

[63]*GET FiT Uganda: Annual Report 2016* (www.GETFiT-reports.com/2016).

of hydropower projects is that when completed they will provide good service and are easier to finance. However, there are several disadvantages:

- The "commenced" date is the date when the actual construction of the hydropower plant begins. Also, considerable time is needed for planning and to obtain all of the permits, etc.
- A considerable number of unforeseen difficulties may occur during the construction.
- The construction can face difficulties in a country such as Uganda, where the infrastructure at many places does not exist.
- The establishment of the water reservoir could submerge inhabited areas. Hence, the residents would need to be relocated.
- The capacity of a hydropower plant is weather dependent.

Table 4.2 List and status of the GET FiT hydropower projects (end of 2016)

Project	Capacity (MW)	Cost (million USD)	US$/MW (million)	Commenced	Expected completion date
SITI I	6.1	14.8	2.43	March 2015	March 2017
MUVUMBE	6.5	14.1	2.15	September 2015	March 2017
RWIMI	5.5	20.8	3.78	July 2015	No date
NYAMWAMBA	9.2	26.8	2.91	Q4 2015	Q1 2018
LUBILIA	5.4	18.7	3.46	March 2016	Q4 2017
WAKI	4.8	18.1	3.77	May 2015	Q4 2017
SITI II	16.0	33.0	2.06	August 2016	Q3 2018
SINDILA	5.0	17.0	3.40	August 2016	Q3 2018
KYAMBURA	7.6	24.0	3.16	Q1 2017	Q4 2018

4.6.2 Cogeneration (Biomass: Bagasse from Sugar Production)

Kakira Sugar Company is the largest sugar manufacturer in Uganda. The byproducts of the sugar manufacture from sugarcane are the crushed sugarcane (bagasse), which is burned to produce hot water and steam. The hot water is used in sugar production and the steam is used to produce electricity. Kakira Sugar Company participated in the GET FiT program and increased its electricity generation by 20 MW (Table 4.3). Now the cogeneration meets all the energy requirements of the Kakira factory complex plus the town on a 24-hour basis as well as provision for the national grid.

Table 4.3 GET FiT Kakira cogeneration project

Project	Capacity (MW)	Cost (million USD)	US$/MW (million)	Completion date
Kakira	20	57.0	2.85	Mid-2015

4.6.3 Solar PV Projects

KfW originally had not planned to deploy solar PV systems, but the cost of solar PV systems plummeted in the past few years and investors have shown great interest in solar PV investments in Africa. The great potential and, as we will see, the short lead time and geographic flexibility—as it can be deployed near existing power lines or existing power stations—have lead the ERA to suggest GET FiT to include solar PV projects to be connected to the grid.

4.6.3.1 Soroti solar PV project

The GET FiT management decided in October 2014 to add a solar PV project to the suggested portfolio of projects in Uganda. As solar PV was the first of its kind, the Soroti 10 MW PV project (Figs. 4.4 and 4.5 and Tables 4.2–4.4) faced a number of regulatory and financial challenges; therefore, the financial close was achieved only in January 2016.

From that point, the GET FiT management received several surprises. The first surprise was the project lead time. The contractor, Spanish company TSK,[64] started construction immediately in February 2016, and nine months later on November 24, 2016, the Soroti 10 MW project started its commercial operation (delivering 17 GWh each year). It could

---

[64]http://www.grupotsk.es/.

have been achieved in five months had the developer not experienced external delays due to customs clearance and damaged equipment during shipping (transformers). The contractor used a local work force of about 120 people, including engineers.

Figure 4.5 Soroti 10 MW solar PV power plant,[65] East Africa's largest Solar PV plant (2016).

Soroti's 10 MW solar PV system was the first project in the 3 years of KfW GET FiT operation which was successfully completed (the Kakira bagasse system was only an addition to an existing system).

Table 4.4 GET FiT Soroti solar PV projects

Project	Capacity (MW)	Cost (million USD)	US$/MW (million)	Commenced	Expected completion date
Soroti	10	19	1.9	February 2016	November 2016
Tororo	10	19.6	1.96	December 2016	October 2017

4.6.3.2 Tororo solar PV project

Based on the success of the Soroti Solar PV project, GET FiT launched another solar PV project, Tororo, also a 10 MW solar

[65]www.GETFiTs-reports.com/2016.

project (Fig. 4.4 and Table 4.4). Construction started in late 2016 and the project is expected to be completed in Q3 2017.

4.6.4   Wind Energy Projects

Interestingly, GET FiT as of today has not planned to install any wind energy project.

4.6.5   Conclusion

The REFiT program produces huge market for PV and CSP in countries where FiT provides sufficient profit and investors are well protected. In Africa, there are many countries where this is not the case. KfW's pilot program indicated that in these countries, the GET FiT systems work excellently. However, it also indicated that in these countries PV systems are preferred. This is described in *GET FiT Uganda—Annual Report 2016*:

On the first page, **Eng. Ziria Tibalwa Waako**, ERA chief executive officer, writes:

> *On December 12, 2016, the Ugandan Electricity Supply Industry through GET FiT support achieved yet another major milestone of commissioning the first grid connected Solar Photovoltaic Plant in the country.*

On the second page, **Kristian Schmidt,** EU ambassador to Uganda writes:

> *On the 12th of December 2016, we celebrated the inauguration of the first GET FiT solar plant in Soroti—with a production capacity of 10 MW the biggest in Eastern Africa....*

Another 10 MW Solar Plant will soon be constructed in Tororo supported by the EU and the GET FiT program.

On December 12, 2016, the inauguration date of the Soroti 10 MW solar PV system, KfW realized that solar PV and not hydropower is the most important RE energy source in the GET FiT program. This became evident in Uganda and will be followed in other countries in the African continent. This opened up an unexpected large solar PV market.

4.6.6 The Future of the GET FiT Program

The success and the experiences of the GET FiT program in Uganda opened the potential for the rollout of the program to other countries in the Sub-Sahara region.

4.6.6.1 Zambia

The Zambian government is in the process of finalizing its REFiT program and rules, which is expected to be completed in 2017.

KfW advanced in 2016 the preparation to launch the GET FiT program as soon as the Zambian REFiT rules are established. In December 2016, the German government committed full funding for Phase I of the program. Learning from the GET FiT program in Uganda, it is planned that a Solar PV system will be the first project. A reverse-auction is planned to be conducted for the acquisition of a 50 MW capacity PV system In 2017. The launch date is, however, dependent on the formal adoption of the REFiT program by the Zambian government. Phase II of Zambia's preparation for small hydro- and biomass systems is planned to be initiated in 2017.

4.6.6.2 Namibia

Namibia at present has one 347 MW hydropower and one 120 MW (since 2015, only 30 MW is in operation) coal-powered station and also several diesel-powered units. Several large hydropower stations are in the planning stages.

Namibia is planning to add 171 MW of renewable electricity to the national grid, which is about 25% of the demand. Small-scale RE producers are expected to play a leading role. Therefore, the REFiT program was initiated by the Electricity Control Board (ECB) and was finalized in October 2015. This interim program is for 70 MW of solar PV. The program limits the production capacity to 5 MW. Therefore, 14 projects are commissioned. The Namibian Power Company (NamPower) prepared the PPA for biomass, concentrated solar power, solar PV, and wind systems (interestingly, hydropower was not included).

Recently, net metering was also approved, which resulted in many private buildings adding solar PV roofs.

KfW GET FiT program in Namibia

KfW was informed that the REFiT and the net metering programs are working very well in Namibia for solar PV. On the other hand, the northern part of Namibia—an area equivalent of the size of Germany, Austria, and Switzerland combined—is being taken over by encroacher bushes and trees. The Namibian government has requested KfW to undertake a detailed design and implementation readiness study to develop a program concept for a GET FiT program "bush-to-electricity" in Namibia. The study shall be financed by the government of Germany. It will be carried out in the course of 2017 (see also Section 8).

4.6.6.3 Mozambique

In 2014, the government of the Republic of Mozambique has introduced a Renewable Energy Feed-in Tariff (REFiT) Regulation to attract private investments to the Mozambican energy sector. Private investments for the same reasons as in Uganda have not materialized as expected. In order to make the Mozambique REFiT program operational, the government of Mozambique requested that KfW undertake a detailed design and implementation readiness study to formulate the GET FiT concept in Mozambique. The governments of the United Kingdom and Northern Ireland will finance the study. This program will be carried out in 2017.

4.7 Deserts as a Source of Electricity

Desertec may be a classic example of the cart getting ahead of the horse. The Desertec concept was conceived originally by German physicist Gerhard Knies in 1986 as he was searching for a clean energy alternative to nuclear power in the wake of the Chernobyl nuclear accident. He recognized that "...in just six hours the world's deserts receive more energy from the sun than humankind consumes in a year " and proposed a system for generating electricity in sunshine-rich North Africa using concentrating solar power (CSP) plants (see Chapter 2.2) and sharing that energy with Europe via a cable under the Mediterranean. The revenue from the sale of that electricity could then be used as a "cash crop" for North Africa. In 1999, the International Energy Agency (IEA) established a new group, PVPS-Task 8, to study the possibility of large-scale generation of electricity using solar PV in desert areas. Task 8 subsequently issued a series of technical reports, including its final report in 2015, entitled *Energy from the Desert: Very Large Scale PV Power Plants for Shifting to Renewable Energy Future*. It concluded that "...desert areas have abundant potential for PV power plants.... The potential annual generation by PV power plants within suitable desert area is...approximately 5 times of the world energy demand and 33 times of world electricity generation in 2012."

The concept was then developed further by TREC, a group of scientists and other experts with an interest in clean energy, and expanded to include a broader range of renewable energy technologies.

In January 2009, TREC evolved into the non-profit Desertec Foundation, which set as its goal the promotion and implementation of its clean energy concept for all the world's deserts. Later, in October that year, to further support the concept, the Foundation and 12 European companies created a largely German-led partnership company called Desertec Industrial Initiative (Dii) GmbH. Its initial purpose was not to build commercial power plants and transmission lines but to carry out studies and develop frameworks related to technical feasibility and financing. Its longer-term goal was to translate the

Desertec concept into a profitable business project that would deliver 20% of Europe's electricity by 2050.

Three foundational studies, funded by the German Federal Ministry for the Environment, Nature Conservation, and Nuclear Safety, were carried out between 2004 and 2007. They evaluated the potential for renewable energy in the Middle East and North Africa (MENA), the potential for an integrated electric power transmission grid linking MENA and Europe, and the potential for producing fresh water in conjunction with CSP generation. They concluded that because of the high solar insolation levels in MENA, CSP plants in the MENA desert regions were more economical for Europe than the same kind of plants in southern Europe, despite transmission losses in the connecting cables.

This conclusion is based on the fact that the Sahara desert is the world's sunniest area year-round and the efficiency of modern transmission systems is high. It is a large area (more than 3 million square miles) that receives, on an average, 3,600 hours of sunshine yearly and in some areas 4,000 hours. This translates into solar insolation levels of 2,500–3,000 kWh per square meter per year. A fraction of the Desert's area could generate the globe's entire electrical demand.

It was also noted that the Sahara desert is one of the windiest areas on the planet, especially on the west coast. Average annual wind speeds at ground level exceed 5 meters per second in most of the Desert and reach 8–9 meters per second in the western coastal regions. Wind speeds also increase with height above the ground, and the Sahara winds are quite steady throughout the year. This is critical for the economics of wind energy as the energy derived from a wind turbine scales at the third power of the speed of the air passing through its blades.

A high-voltage direct current (HVDC) transmission system was proposed to carry the electricity under the Mediterranean to nearby points in Europe. HVDC is a well-established efficient method for transmitting large amounts of power over long distances. This electricity would then be distributed throughout Europe via the European grid connecting its various countries. With the advent of modern control electronics, this grid would evolve into a smart "supergrid" that would tap into electricity from MENA and other generating sources in Europe and share power as needed.

This utopian concept had a predecessor, the Atlantropa project, which in the 1920s proposed that a giant hydropower dam be built across the Strait of Gibraltar. The energy available from this dam would then provide large amounts of hydroelectricity to the whole Mediterranean region and facilitate integration of the electricity systems of Europe and North Africa. It reached great popularity in the late 1920s and early 1930s, and again for a short period after World War II, but was forgotten with the death of its principal advocate in 1952.

A few years after it was established, Dii GmbH ran into serious reality and political problems. Solar PV prices started coming down dramatically, countries began installing increasing amounts of PV and wind, and it soon became clear that Europe could provide for its own renewable energy needs on its own. Power generation in North African deserts was also beginning to be seen as a means of satisfying growing demand for electricity from Africa's rapidly increasing population. As a result, Desertec attracted very little funding, companies began dropping out of the project, and Dii was forced to revisit its original strategy.

Dii also ran into a problem related to power transmission. Spain, an obvious connection point for the cable from North Africa, already was struggling with too little transmission capability to the rest of Europe for its own renewable energy production, and Europe more generally was struggling in a similar manner. Susanne Nies, head of Energy Policy and Generation at Euroelectric, the European electricity trade association, summarized the situation concisely when she stated: "At a very basic level, we are still missing lines and capacity for export.... It is difficult to argue that the European Union needs the additional renewable energy capacity...additional imports from third countries would certainly compound the problem."

Another area of concern is that Dii is seen by some as just another "colonial scheme" for exploiting the riches of Africa for the benefit of previously colonial European nations. In 2011, Daniel Ayuk Mbi Egbe of the African Network for Solar Energy put it succinctly: "Many Africans are skeptical about Desertec. Europeans make promises, but at the end of the day, they bring their engineers, they bring their equipment, and they go. It's a new form of resource exploitation, just like in the past." Mansour Cherni of Tunisia made a similar point at the World

Social Forum in 2013 when he asked: "Where will the energy produced here be used? Where will the water come from that will cool the solar power plants? And what do the locals get from it all?"

Nevertheless, Dii still has it supporters, including the shareholders German utility Innogy, Saudi Arabia's Acwa Power, China's State Grid, and countries like Morocco, with large solar resources, that have always been supportive of Desertec and Dii. Saudi Arabia recognizes that their ambitious solar plans may benefit from connecting their electricity grid to the west, so they could export some of the solar power they don't need instantly to their neighbors' neighbors.

All these considerations led Dii to reset its goal from bringing electricity from North Africa to Europe, to "...creating integrated markets in which renewable energy will bring its advantages...." An important step was to move the company from Munich to Dubai and to focus on helping governments and companies implement projects. Nonetheless, there has been a spate of articles stating that Desertec had died. And then, in August 2017, an article appeared in the e-journal *CleanTechnica* entitled "The Desertec Sahara Solar Dream Didn't Die After All—It's Baaaack...." It reported that Nur Energie, a London-based energy company that specializes in CSP technology, had filed "...a plan with the Tunisian government to export 4.5 GW of solar power from the northeastern edge of the Sahara to Europe." Its rationale seems to be that CSP technology has matured in recent years and that the 2015 Paris Agreement on Climate Change will be an important motivator for deploying CSP. The Nur Energie initiative will certainly be a big test for the Desertec concept, given that CSP is more costly and harder to maintain than solar PV. However, it benefits from an important characteristic of CSP power plants—they can store thermal energy and generate electricity long after the sun has gone down. Many CSP plants already exist, starting with plants in California in the 1980s, including the 100 MW Shams1 plant in Abu Dhabi, and others are under construction, also in the MENA region. The recent Dubai tender results, flowing from their request for proposals for a 200 MW CSP plant to provide night-time electricity, have shown that CSP with integrated thermal energy storage can produce solar electricity for substantially less than 10 cents per kWh, making it a real alternative to fossil fuels and certainly nuclear power. The next few decades will decide if CSP is a long-term solar energy technology and whether the Desertec cart can finally be put behind the horse of renewable energy.

5

Existing and Emerging Solar PV Markets

5.1   Introduction

Chapter 5 discusses many of the practical applications to which solar energy has already been applied. These include pumping of water from underground aquifers, provision of clean water through desalination and decontamination, powering of free-standing telecommunication towers that facilitate wireless telephony and Internet communication (which enables remote education and medical care capabilities), and the growing use of solar in the mining industries that are prevalent in Africa and the Middle East.

The Sun Is Rising in Africa and the Middle East: On the Road to a Solar Energy Future
Peter F. Varadi, Frank Wouters, and Allan R. Hoffman
Copyright © 2018 Pan Stanford Publishing Pte. Ltd.
ISBN 978-981-4774-89-5 (Paperback), 978-1-351-00732-0 (eBook)
www.panstanford.com

5.2   Water Pumping Utilizing Solar Electricity

Water pumping using solar electricity is now and for the future extremely important for Africa and the Middle East.

The first experimental solar-powered water pump was installed on the island of Corsica in 1974. The experiment was initiated by Wolfgang Palz of the European Commission. It was implemented by Pompes Guinard, a French company, and the University of Lyon under the direction of Dominique Campana.

It is, however, not widely known that the first solar water-pumping project in 1977 was in Africa. Georges Chavanes[66] was the president and Paul Barry the managing director of Leroy-Somer, a subsidiary of which was Pompes Guinard, a manufacturer of pumps. They learned from the success of the Guinard solar water pump installed in Corsica and believed that not only solar water pumps could become a business, but they could install them in the western part of Africa and help the poor people there. They chose the western part of Africa for the first solar water-pumping project because the countries in the region were ex-French colonies with close connections to France and, more important, had a large poor population that did not have access to sufficient potable water for themselves and their animals and water for agriculture. The shortage of water was despite that an extremely large amount of ground water was available just a few meters below the surface. The area had no electricity and therefore water could not be obtained by pumping. Guinard developed pumps suitable for operation with PV electricity and installed many pumps in that area. The project was very successful.

The Egyptian government was studying desert reclamation for agricultural purposes. One requirement was proper soil. It was established that the soil (sand) in the Egyptian Western Desert called East Owainat had the ingredients that make it suitable for agriculture. The other requirement was water. The research in the late 1970s by the distinguished Egyptian scientist, Dr. Farouk El-Baz, director of the Center for Remote Sensing at Boston University in Boston, Massachusetts, who was also the scientific advisor to the late President Anwar Sadat, indicated that below East Owainat is a large water reservoir. The experimental farm site in the East-Owainat desert was finally selected, because

---

[66]Georges Chavanes became the Minister of Commerce of France (1986–1988).

the Egyptian oil company General Petroleum Company (GPC) while drilling for oil or gas found a large amount of water only 25 m (80 ft.) below the surface as predicted by Dr. Farouk El-Baz. Today we know that it is the Nubian Sandstone Aquifer[67] System, estimated to contain about 400,000 km^3 of water, which is the equivalent of 4000 years of the Nile river's flow and could provide a plentiful supply of water.

Figure 5.1 Utilization of a solar water pump to provide drinking water.

The success of such reclamation projects rested upon developing efficient, economical means to bring the plentiful underground water to the surface. But for the pumping of water, electricity was needed. The nearest point from where an electrical line could be brought to this area was more than 800 km away. To establish the 800 km connection for an experimental project was too expensive. Utilizing diesel generators to provide electricity for water pumping and irrigation and for the people working at the farm was also impossible, because the infrastructure of suitable roads to truck the fuel for the diesel generators was non-existent at that time. An automobile trip

---

[67]An aquifer is an underground layer of water-bearing permeable rock or unconsolidated materials (gravel, sand, or silt) from which groundwater can be extracted using a water well.

from Cairo to East Owainat in those days required 3 days on "roads."

To verify that the required amount of water can be successfully produced from the aquifer for a 10 acre experimental farm and that the sand quality and the climate in East Owainat is good for agriculture, the Egyptian government decided to provide electricity by using solar energy, which was found successful in the Pompes Guinard program on the west coast of Africa.

One of the first large-scale solar water-pumping systems for agricultural purposes in the "real" Sahara (which is not close to inhabited areas) was the one Solarex installed in East Owainat in the southwest corner of Egypt, close to Libya in the west and Sudan in the south where the Egyptian government wanted to study desert reclamation.

To supply electricity for the entire farm, including irrigation, Solarex designed and installed in 1984 a PV system (see Fig. 5.2) to deliver 350 m^3 of water a day during the summer season (approximately 35 m^3 an hour over a 10-hour duty day). In addition, a 350 m^3 water storage tank was provided. The installed pumping system was powered by an array of PV modules producing 21.6 kW at peak power. It also included a sizeable battery bank (352 kWh) to provide electricity for lights and irrigation at night.

Figure 5.2 Owainat, Egypt, PV water-pumping system (1984).

The experimental project was extremely successful. In light of the result, reclamation of 230,000 acres to serve 100,000 people and create 20,000 jobs was planned. As of 2013, 110,675 acres have been reclaimed in the project for agriculture use. Almost 35 years later, the Egypt State Information Service advertised

East Owainat as follows: "The East Owainat project has turned the barren desert into a green paradise in which trees, flowers and fruits have grown for changing the face of the Egyptian history."

These and other small and larger—we can call them—experimental solar water-pumping applications were so successful that in 1990 the European Union (EU) initiated a large-scale water-pumping project in the semi-arid Sahel region part of Africa. The Sahel spans Africa from the Atlantic Ocean in the west to the Red Sea in the east, a 5400 km-long belt that varies from several hundred to a thousand kilometers in width, covering an area of 3,053,200 km². It is a land area almost half of the size of the contiguous U.S.A. It consists of semi-arid grassland, savannas, and steppes between the wooded Sudan savanna to the south and the Sahara to the north.

The Sahel region comprises the West African countries of Burkina Faso, Chad, The Gambia, Guinea Bissau, Mali, Mauritania, Niger, Senegal, Sudan, and Eritrea. At least 135 million people live in the region, with some 30% in urban areas, according to the Permanent Inter-State Committee for Drought Control in the Sahel (French abbreviation, CILSS).

The water table under the Sahel is at most 100 mbgl[68] and in some areas is as close to the surface as 2 mbgl. Despite the rainfall in the Sahel area being only between 150 and 600 mm/year, this water table is practically independent of the amount of rain and it seems to be rising and not dropping. This means that the pumping of water by PV is very much feasible.

A large part of this population needs to transport water from a distance. The average African women and young girls spend a considerable amount of time carrying up to 60 liters (60 kg) of water 10 km every day.[69] The EU implemented a Regional Solar Program (RSP) by installing a large number of solar water-pumping systems. The RSP is a vast, ambitious, and innovative program that was launched in 1986 by the Heads of States of the CILSS countries and financed by the EU. The importance of water can be seen from the information[70] that in 2008 39% of the inhabitants

---

[68]"meter below ground level."

[69]Dominique Campana (2011). *PV Power Systems for Lifting Women Out of Poverty in Sub-Saharan Africa*, in Wolfgang Palz, ed., *Power for the World*, Pan Sanford Publishing, Singapore.

[70]https://www.unicef.org/wcaro/overview_4552.html

of the Sahel do not have access to drinking water (more than 50% in Burkina Faso, Guinea Bissau, and Mauritania). The EU's RSP-1 was carried out from 1990 to 1998 and produced excellent results. It was continued by RSP-2, which was completed in 2009 and provided more than a million people with drinking water. Under RSP-2, 1,000 solar water pumps as well as 16 pumping systems for irrigation were established in the Sahel region.

Based on the previous experience, the EU solved the problem of the after-installation maintenance of solar water-pumping systems. What unfortunately was not easy to solve was the theft of a substantial number of solar modules (up to 30% in some of the countries) and inverters from the pumping sites. Each country or region developed their own strategies and introduced safety measures. The stealing of PV modules indicated that there was a great need to provide electricity for other purposes, too. Most likely, the stolen PV modules were used for lighting and powering radios or TV receivers.

The attractiveness to steal solar modules and inverters changed drastically after the beginning of the 2000s and was nil by the end of the RSP program. The stealing of solar modules became not a lucrative business anymore because the price of solar modules decreased very much. In 2000, the price of solar modules was \$3.90/W. In 2017, it became less than \$0.30/W. Also, the panels used for pumping typically have higher voltages these days, making them unsuitable for home battery charging. The global sale of pumps used for solar water pumping became so large that pump manufacturers started to make water pumps specifically for solar applications, and to reduce the cost of solar pumping, DC instead of AC motors were used to avoid the cost of the inverter. Therefore, there was no question of inverters getting stolen, because there were no inverters in solar pump systems anymore.

Pump manufacturers developed their "solar pump" lines. Some of the pump manufacturers switched their entire production to manufacture only solar pumps. Solar pumps can now directly be connected to PV module systems. At present, two types of solar pumps are available: "submergible pumps" and "surface pumps."

- "Submergible pumps" are designed to be lowered into boreholes to pump water from a depth. Some of these pumps are able to pump water from as deep as 450 meters.

- "Surface pumps" are used for shallow wells, ponds, streams, or storage tanks.

At least 30 solar water pump manufacturers exist in the world. Their products can be purchased on the Internet, but several have sales representation in practically every country in Africa and the Middle East and they are also installing the solar water pumps for their customers.

In many applications, solar water pumps are connected to a water tank (see Figs. 5.1 and 5.2) to make water available around the clock. The pumped water is mostly used for drinking by people and animals and for irrigation of high-value crops. In Africa and the Middle East, several ten-thousands of solar water pumps are installed annually, the great majority of which are submergible.

That so many water pumps are installed yearly—and the above number is probably a low estimate—is because North Africa and the Middle East are considered mostly dry area. In Africa, at least 65% of the entire land area is considered to be either arid, where the yearly rainfall is around 50 mm/year, or semi-arid with 150 mm precipitation per year. In the Middle East, most of the land is arid or semi-arid. What is not well known is that an immense amount of underground water exists very close to the surface of those arid and semi-arid areas in Africa and the Middle East. One of these, as already mentioned, was under Owainat in Egypt. To utilize that large amount of water for drinking and irrigation, pumps are needed. The problem is that in most of those regions, there is no supply of electricity to operate the pumps. The solution, as it was in the Owainat area, is to use solar electricity to pump the underground water.

The following is a review of the vast area of Africa and the Middle East to see where solar water pumping would help the inhabitants to solve the problem obtaining water for themselves, for animals, and to establish land like Owainat did for agriculture.

5.2.1 Africa

(A) The vertical length of the African continent is about 7250 km. The availability of rain water on the African continent is vastly different in large horizontal bands.

(B) On the northern part of the continent is a 1900 km-wide extremely arid band, the largest part of which is called the Sahara. On the northwestern part of this band is a very short and narrow area, about 320 km wide and 560 km long, that receives about 400 mm precipitation annually.

(C) On the south of the Sahara is an average 650 km-wide semi-arid band called the Sahel.

(D) On the south of the Sahel is a 2550 km-wide band that receives lots of precipitation, around 1,000 mm annually.

(E) To the south of the region mentioned in point C until the tip of Africa is a band about 2,000 km wide. The west side of it is arid, while the rest of it has semi-arid and wet area.

The region in C and part of D receives rainwater and has many large lakes and rivers sufficient for people, animals, and farming. One has to realize that not all of this water is suitable for drinking. Natural groundwater levels are approximately 0 to 25 mbgl. Near lakes or rivers, it is <7 mbgl. This water can be pumped utilizing an electrically operated pump and many times even with a hand pump. In many places, electricity is not available, but solar pumps can be used.

The majority of the A, B, and part of C are arid or semi-arid. This area has no rain or river water, but as explained above, in many areas there are vast reservoirs of ground water called aquifers. Those areas, which are considered as water scarce, have substantial amount of water reserves underground as we have seen from Egypt's Owainat example.

In fact, aquifers and underground water represent a water resource that is so immense that it is different in magnitude compared with all freshwater sources in Africa.[71] The water stored in this underground storage media is estimated to have about 30 million times more water than the annual rainfall and the water in all lakes in Arica. The estimated groundwater storage in 19 countries in Africa ranges from 7,000 to over 100,000 km³. These underground sources of water in most cases are close to the surface. For example, as mentioned previously, in the Owainat area of Egypt, the aquifer's water level was at 20 meters.

---

[71]http://iopscience.iop.org/article/10.1088/1748-9326/7/2/024009.

A study[72] indicates that the majority (85%) of Africa's population lives in the regions where the depth of groundwater is 0–50 mbgl and 8% lives where the depth of the groundwater is 50 to 100 mbgl.

This means that people living in most parts of Africa are able to obtain sufficient water needed for drinking and also for irrigation from either surface or underground. Some of the underground water can be pumped with a hand pump, but in the majority of cases, the water level is deeper and requires an electric pump to lift the water. As discussed in this book, the majority of Africa has no access to electricity and has now started to use solar water pumps. That is why a network of solar pump distributors exists practically in every country in Africa and at least tens of thousands of solar water pumps, a great numbers of those submergible, are being installed every year.

5.2.2 Middle East

The Middle East, also called West Asia, is a large land area. For the possible utilization of solar water pumping, we will discuss only a section of it, which is east of Iran and south of Turkey. This entire area is either semi-arid or arid. With respect to the availability of water, this area can be divided into three segments. The northern part is Iraq, the western segment consists of Syria, Lebanon, Jordan, Israel, and the Palestine Authority, and finally Saudi Arabia and its neighboring countries in the east and the south.

Iraq

The eastern part of Iraq where the rivers Tigris and Euphrates flow was in the ancient times the cradle of civilization. Although Iraq has these two major rivers and several smaller rivers and lakes, it is facing a shortage of water. Solving this problem will require groundwater utilization. In addition to the groundwater that naturally occurs around the rivers and lakes under the southeastern part of Iraq, there is the Wasia-Biyadh-Aruma aquifer system (North). The size of this aquifer is estimated to be about 87,000 km^2, of which 35,000 km^2 is in Iraq and 52,000 km^2 is in Saudi Arabia. The depth to the water level of this aquifer is

[72]H. C. Bonsor, A. M. MacDonald. An initial estimate of depth to groundwater across Africa, British Geological Survey, Groundwater Science Program OR/11/067.

about 200 mbgl and the soil in this area is fertile. It is expected that solar-powered pumps will be widely utilized in the eastern area where no electricity is available. In the large desert area underneath which the Wasia-Biyadh-Aruma aquifer system lies, submergible solar water pumps will be used.

Western segment

This semi-arid area consists of Syria, Lebanon, Jordan, Israel, and the Palestine Authority. The available surface water is not sufficient for them, but the area has several underground aquifers providing the additional needed water at present. "Present" means that with the increasing demand for water, it will become a very serious problem. Solar water pumping utilizing submergible solar water pumps may be helpful in certain areas such as the Negev desert of Israel, but in the rest of this area the problem is not the availability of electricity like in Africa but the availability of water.

The east and south area of the Middle East

This area consists of Saudi Arabia and its neighboring countries on its east and south. The entire area is arid. About 50% of the fresh water needs of Saudi Arabia are met from the desalination of seawater. The country has 30 desalination plants, which account for about 20% of Saudi Arabia's total energy consumption. Riyadh, the capital, is supplied with desalinated water, which is pumped about 500 km from the Arabian Gulf. This pipeline has four pumping stations, as the water has to be pumped from the desalination station at the sea level to the Riyadh High Point Terminal at 700 m. Each pumping station has 20–30 MW capacity.

Saudi Arabia has a 5,390 km network of water transmission pipelines to transport water in bulk from the sea-level desalination plants (more than 4.6 million m^3/day water installed capacity) to the major consumption centers, some of them located at high elevation such as Riyadh. The system has a total of 46 pumping stations. Substantial electricity is required for the pump stations designed to accomplish a 4.4 m^3/sec flow rate at very high pressure. The installation of PV electricity-generating systems for the pumping stations would substantially reduce the daytime need for the very expensive oil-produced electricity.

This part of the world is well known for its huge quantity of oil and gas production and its extremely large underground oil and gas reserves. It is, however, not widely known that a very large number of aquifers exist in a vast area under the Arabian Peninsula. One of the largest aquifers is the Wasia-Biyadh-Aruma system (south), the continuation of the Wasia-Biyadh-Aruma System (North). This aquifer extends 2,400 km from Iraq to the southern coast of Yemen with widths varying from 350 to 1,400 km. Many other smaller aquifers also exist and 45% of the country's fresh water comes from this ground water and only about 5% from surface water sources.

Ground water utilization is very much feasible in almost the entire Arabian Peninsula area. However, the risk of its over usage was demonstrated in Saudi Arabia's very ambitious project when, starting in the 1970s, it was encouraged to transform the desert into lush farmland by letting farmers dig as many and as deep wells as they wanted. It was very successful and Saudi Arabia became the world's sixth largest exporter of wheat. As a result, however, they exhausted the aquifer in that area. The farmers' wells and the oases, which had existed since ancient times, dried out. The country then decided that the 2016 wheat harvest would be the last one. According to one estimate, the amount of water in the local aquifer before the farming project started was 500 billion cubic meters, equivalent to the amount of water in Lake Erie, and by 2012, about 80% of that water had been extracted.

The successful utilization of ground water was demonstrated by that farming project. As almost the entire area is above an aquifer but the majority of the area has no connection to the electricity grid, the utilization of PV-powered submergible pumps to draw water has become very popular. One needs no better evidence to prove this than the Internet. On the Web, one can find distributors and/or installers of "solar water pumps" in several countries and cities of the Arabian Peninsula.

5.3   Solar Energy and Clean Water

As discussed in Chapter 5.2, solar energy can play an important role in facilitating access to clean water in rural areas of the developing world. These areas are usually too isolated for on-grid water pipe infrastructure to be built, and the responsibility for obtaining clean water is on the women and children in the villages. They often have to spend many hours a day walking long distances to fetch water, which is often contaminated.

The close connection between water and energy, now often referred to as the water–energy nexus, has been explicitly recognized and carefully described in recent years. The large amount of solar energy that falls on Africa and the Middle East, both regions with serious water problems, offers a unique opportunity to put that energy to good use in providing clean water for drinking, sanitation, and agriculture. The means by which solar energy can do so is discussed in the following paragraphs.

5.3.1   Desalination

Desalination is the process by which dissolved salts are removed from salty (saline) water to produce potable water. Saline water is characterized by its salt concentration in parts per million/ppm: highly saline (10,000–35,000 ppm), brackish (1,000–10,000 ppm), and fresh water (less than 1,000 ppm). Ocean (sea) water, which accounts for 97% of the water on the earth, is highly saline. Brackish water can be found underground in fossil (ancient) water aquifers, and is created when seawater invades fresh water supplies.

There are quite a few desalination technologies available today, starting with significant advances in the early 1900s and advancing rapidly during World War II, when it became necessary to provide potable water to military personnel operating in remote dry areas. It became commercially viable in the 1990s.

The most common type of desalination in use today is reverse osmosis (RO), which accounts for two-thirds of installed capacity. It requires no thermal energy to evaporate saline water that is then condensed free of its salt (the historic and traditional method, called distillation), just mechanical pressure

(800–900 psi) to force salty water through a membrane that separates the salt from the water. This creates the opportunity for large-scale use of solar energy to power desalination as electricity is needed to power the pumps that create the needed pressure. RO is the most energy efficient of all the desalination technologies, requiring 3.0–5.5 kWh per cubic meter of fresh water produced, depending on the salinity of the source.

Desalination is important in a water-rich world (the earth contains an estimated 300 million cubic miles of water, with each cubic mile containing more than one trillion gallons) because regular ingestion of highly saline water can kill people and animals. An Arab saying tells us "water is life" but that water has to be potable.

Solar energy is important for the delivery of clean water services in three ways: It can be used for pumping water from underground aquifers (see Chapter 5.2), it can be used to power the pumps that push saline water through RO membranes, as is discussed below, and to disinfect contaminated water. The third application is discussed in Chapter 5.3.2.

The availability of vast quantities of solar insolation in Africa and the Middle East, both of which include many of the world's most water-scarce countries, makes it a natural energy source for desalination, especially as solar energy costs have plummeted in recent years. As a result, many solar-powered RO units are being deployed in both regions, with many more planned. Today there are more than 18,000 desalination plants operated worldwide, producing approximately 90 million m^3 of fresh water daily for 300 million people.

Large-scale solar-powered desalination was discussed extensively at the 2013 World Future Energy Summit in Abu Dhabi. Further, in 2015, Saudi Arabia, which depends on the desalination of seawater for 60% of its fresh water supply, announced that a 15 MW solar array would be used to supply 60,000 m^3 of desalinated seawater daily to the city of Al Khafji. The array is part of the Mohammed bin Rashid Al Maktoum solar park, which is scheduled to become, at 1,000 MW, the largest solar power plant in the Middle East in 2030. Many other countries are also examining the potential of solar-powered desalination, including Abu Dhabi, Kuwait, Qatar, and even the United States. A 400 KW concentrating solar power system (parabolic troughs) is being tested in California's Central Valley to desalinate and reuse agricultural runoff and other types of contaminated water, directly using the system's collected thermal energy to drive distillation.

5.3.2 Disinfection

Solar-powered disinfection uses solar-generated electricity to drive systems that remove bacteria, viruses, protozoa, and worms from contaminated water, thus making it safe to drink. One procedure, widely used in Europe in medium to large disinfection facilities, is to create O_3, an unstable form of oxygen, by an electrical discharge in air or pure oxygen. The unstable O_3 molecules then attach to, oxidize, and destroy the contaminating pathogens.

Another way of killing pathogens is to expose them to ultraviolet radiation, which disrupts their DNA and prevents their reproduction. This concept was developed by Dr. Ashok Gadgil of Lawrence Berkeley National Laboratory, as a means to control the outbreak of cholera and other diseases with a technology that would be inexpensive and easily maintained without a skilled operator. It is well known that UV radiation in the wavelength range 240–280 nm has herbicidal effect, and recent research has pinpointed 260 nm as the most biologically effective wavelength.

The system developed by Dr. Gadgil, called UVWaterworks, and now commercialized by International Health, Inc., can disinfect about one ton of water per hour at a cost of about 5 U.S. cents. It works by passing unpressurized water under a UV lamp, which does not come in physical contact with the water. The lamp, typically with a broad output spectrum but designed to put most of its radiation into the desired wavelength region, can be powered by a single solar PV panel. More than a million units are now in operation around the world.

An important related development is that solid-state LED technology has now been extended to the UV region, offering the possibility of UV light sources that can be designed for narrow band emission centered at 260 nm. If they can be produced inexpensively, they will be attractive replacements for the less efficient UV lamps currently on the market.

5.4 Off-Grid Telecom Towers

One of the development success stories has been the phenomenal growth of cellular subscriptions worldwide. There are now more mobile devices in use than people in the world. According to the GSMA, the mobile operators' industry association, Africa currently has over 800 million mobile connections and some 450 million unique subscribers (a unique subscriber is an individual that may have more than one SIM or connection).[73] The coverage of mobile networks varies from 10% to 99% across countries in Africa with an average mobile coverage of 70% in sub-Saharan Africa. Given that two-thirds of the population in sub-Saharan Africa does not have access to grid electricity, there are two electrification issues connected to mobile communication: charging the mobile phones and powering the telecom towers that relay signals (Fig. 5.3). How small PV systems can charge mobile phones and how mini-grids and pay-as-you-go business models can help the dissemination of PV electricity is dealt with elsewhere in this book. This chapter deals with the telecom towers providing the mobile coverage throughout Africa.

5.4.1 Off-Grid or Bad-Grid?

The mobile industry faces challenges powering the growth in Africa. Their customers are spread thinly over the vast continent and need coverage, also in areas where the grid does not reach. However, even in grid-connected areas, the power is often unreliable and telecom towers require back-up power systems. Most of the telecom towers in Africa are in off-grid and bad-grid locations, whereby a bad-grid location is characterized by regular and sustained blackouts. In 2014, sub-Saharan Africa had a total of over 240,000 towers, with 145,000 at off-grid sites. The number of bad-grid sites was estimated at 84,000; so only 5% of the sites can be considered sufficient, from an energy point of view. Most off-grid and bad-grid sites have diesel generators and diesel expenses amount to 40% of the running costs of a typical African tower site. An off-grid site consumes nearly

---

[73]*Tower Power Africa: Energy Challenges and Opportunities for the Mobile Industry in Africa.* http://www.gsma.com/mobilefordevelopment/wp-content/uploads/2014/11/Africa-Market-Report-GPM-final.pdf.

13,000 liters of diesel per annum, while a bad-grid site burns through 6,700 liters. On top of this, one should add 10–15% cost due to pilferage of diesel, which is common and hard to avoid in many countries across Africa.

Figure 5.3 Telecom tower.

5.4.2 Tower operators

Most African mobile network operators (MNOs) expanded their businesses aggressively in the period between 1990 and 2010. Operators are more and more focused on acquiring customers and expanding services, and started outsourcing capital-intensive parts of the business such as the towers. The separation of telecom infrastructure assets from retail telecoms, and the inauguration of an independent telecom tower industry, has taken place in the United States and largely in India, but this transformational change is only starting to take root in other markets. Although African MNOs still own 80% of the towers, a growing number of MNOs are selling (part of) their towers to specialized tower companies (Tower Cos). The number of towers in Africa is expected to grow to 325,000 in 2020. The recent aggressive focus of Tower Cos in Africa has led to majority tower portfolios being transferred to Tower Cos as ownership or management contracts. According to GSMA, by 2020, 60% of towers could be owned by Tower Cos. The largest African Tower Co, IHS Towers, operates more than 23,000 towers in Nigeria, Cameroun, Rwanda, Zambia, and Côte d'Ivoire. Over the past couple of years, IHS Towers has invested more than $500 million upgrading their power systems, adding 72 MW of solar PV in Nigeria alone in 2016. In the same year, IHS had reduced its diesel consumption by up to 50% on average across Africa over the past year.

At a diesel price of $1.5/liter (delivered), the annual energy cost amounts to $28,000 per off-grid tower and $22,000 per bad-grid tower.

The low density of Africa's rural areas, the expensive energy supply, and poor infrastructure make the necessary expansion of the tower network more expensive than elsewhere. Furthermore, co-location of more than one MNO on a single tower site is less usual than in other places such as India,[74] which has a larger percentage of Tower Cos, making the tower service relatively more expensive than if cost for the tower service could be shared among MNOs.

5.4.3 Renewable Energy Towers

Africa has abundant sunshine almost everywhere and with the current price levels for solar systems, there is a solid business case to replace the expensive diesel systems with PV battery systems, or expand an existing generator by adding PV as a fuel saver. Nonetheless, to date, fewer than 10,000 of the more than 240,000 African telecom towers use solar energy. There are a few reasons for that. As explained above, most towers are still owned and operated by the MNOs, and their main objective is to invest in the expansion of market access, market share, and services. Building towers is expensive, so they would rather select a low-investment option. Despite high running costs, diesel generators are relatively cheap. There is also an established network of diesel generator suppliers that know how to run,

---

[74]*The Rise of the Tower Business.* https://www.atkearney.com/documents/10192/671578/Rise+of+the+Tower+Business.pdf/027f45c4-91d7-43f9-a0fd-92fe797fc2f3.

operate, and maintain such equipment. And lastly, the rapidly declining cost of solar PV has made this an appealing option only recently, and the awareness about this is slowly starting to trickle down the supply chain. Nonetheless, the business case for solar PV is now so strong that it is only a matter of time before most towers will have a solar PV energy system. Two developments might accelerate this trend: the transfer of the towers from the MNOs to Tower Cos, and the outsourcing of the energy system to an energy services company (ESCO).

5.4.4 Tower ESCOs

The main non-telecom hardware in an off-grid or bad-grid tower is the energy supply system. Since the MNOs only allow minimal downtime (the typical required availability is higher than 99.5%), this requires dedication. The cost in terms of investment and operation and maintenance expenses are substantial, and recently a new business model has emerged, whereby dedicated ESCOs offer guaranteed electricity to tower operators, whether that is an MNO or Tower Co. The ESCO invests in the hardware (diesel generator, PV panels, batteries, and control hardware) and operates the power system for the tower operator. The power is guaranteed through a service level agreement, and Energy Vision, an ESCO based in Mauritius but active throughout the African continent, claims on their website (www.genergy. vision) a guaranteed 99.8% availability based on a Service Level Agreement with penalties for under-performance. Energy Vision also actively promotes clean power, integrating solar panels and wind turbines where feasible.

Figure 5.4 shows how the mobile phone penetration in sub-Saharan Africa has surpassed 80%, far more than access to electricity, water, or sanitation. Since the network is powered by decentralized energy systems, increasingly using solar power, there is an opportunity to expand the energy system and use it to provide power to rural households and businesses. Since a Tower ESCO's business is to do exactly this, albeit for a single user, the spread of the Tower ESCO business model could well be ignition for the long-awaited breakthrough in the renewable energy mini-grid business. As described above, the tower power business model is evolving from single MNO towers to sharing towers. And with third-party ESCOs gaining traction, the power system could be expanded to include renewable energy systems and cater to surrounding communities. This is already happening in India, and there is no reason why this could not be the next wave of electrification in Africa.

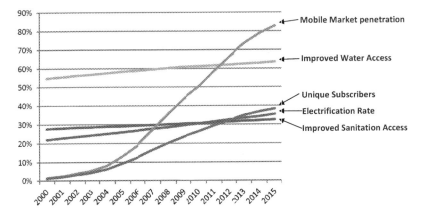

Figure 5.4 Mobile phone penetration vs. energy, water, sanitation access in Sub-Saharan Africa.[75] Source: GSMA, World Bank, IEA 2012.

[75]*Sustainable Energy & Water Access through M2M Connectivity.* http://www.gsma. com/mobilefordevelopment/wp-content/uploads/2013/01/Sustainable-Energy- and-Water-Access-through-M2M-Connectivity.pdf.

5.5 Internet with PV

5.5.1 Internet in Africa

As with electricity, Africa's Internet usage is among the lowest in the world. The following map (Fig. 5.5) shows that the Internet user penetration in most African countries is well below 20%, with the majority even below 10%. Even Egypt and South Africa are below 50%, the only country with more than half the population having Internet access is Morocco, with 56%. To put that in perspective, the top 10 on the global list, nine of which are in Europe, have penetration rates higher than 90%. On top of that, Africans pay 30 to 40 times more for Internet access than their peers in developing countries. Nonetheless, the Internet in Africa is growing rapidly from its current overall penetration level of about 20%. Mobile subscriptions are just shy of 70%, and mobile broadband access accounts for more than 90% of Internet subscriptions.[76]

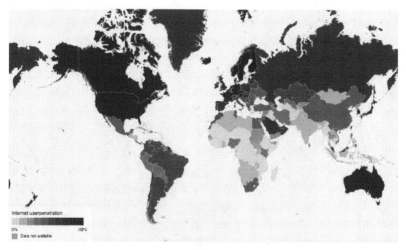

Figure 5.5 Internet user penetration.[76]

According to the World Bank, a 10% increase in broadband correlates to a 1.38% increase in GDP growth, so Internet access

---

[76]Courtesy of Internet Society. www.internetsociety.org.

is a driver for economic development.[77] However, since facilities for Internet access require electricity, the development of the Internet is held back by the lack of easily accessible and competitively priced electricity on the continent. The advent of renewable energy has opened many new possibilities.

In the past decade, several businesses powered by solar energy have sprung up all over Africa. With solar becoming more and more affordable, many businesses have added solar panels and batteries to existing diesel systems. This so-called hybridization reduces fuel cost and also the maintenance requirements for the diesel generator, because its use is limited. However, several new business concepts powered 100% by solar energy have also been developed.

5.5.2   NICE, the Gambia

In 2001, Dave Jongeneele, founder of Better Future,[78] organized the first European-Gambian management exchange program in the Gambia, one of Africa's smallest countries. Better Future takes leaders and their teams on a thought journey that aims at opening their horizon and creates an impact beyond their imagination. Paul van Son, then managing director sustainable energy at Essent, a Dutch utility that is now part of Germany's Innogy, participated in that journey. He discovered that many Gambians already had a mobile phone and walked long distances to charge the battery. There were only few, relatively poorly functioning Internet shops in the larger communities. Water supply and lighting was also a great problem in many locations. There was some basic power grid infrastructure along the coast, but many communities had no or only very expensive access to the grid. The Gambia has among the highest electricity tariffs in the world. Upon his return home, Paul founded the Energy4All Foundation to pave the way for decentralized clean energy, clean water, and communication/Internet services based on small solar units, wind turbines, and batteries. Paul's vision was to build so-called micro-utilities. He quickly found three sponsors:

---

[77]Christine Zhen-Wei Qiang and Carlo M. Rossotto with Kaoru Kimura (2009). Economic impacts of broadband, in *Information and Communications for Development 2009: Extending Reach and Increasing Impact*, The World Bank, p. 35.
[78]https://www.linkedin.com/in/dave-jongeneelen-b7a2a1.

Econcern, Essent, and Rabo Bank. At the time, one of the authors of this book, Frank Wouters, was working for Econcern and got involved. They persuaded Paul to focus on PV and develop an Internet shop concept first as that would lead to quick economic wins. Econcern founded a company called NICE International (**N**ext-Door **I**nternet, **C**ommunication and **E**nergy Service shops), with a subsidiary in the Gambia. Energy4All supported that venture and Wouters was elected the chair of the supervisory board. At a later stage, Schneider Electric and the Dutch development bank FMO joined the sponsors.

NICE developed a concept that would provide Internet and computer services in areas that did not have access to electricity. We started with an extremely efficient design. The selection of all equipment was based on performance and low energy consumption. For instance, NICE's thin client computers used approximately 10 W compared with 200 W for a typical desktop computer. As a result, a complete NICE-center with 30 computers and a cinema (a large flat screen TV) uses the same amount of energy as one air conditioner! A tracker was designed that had eight multicrystalline PV modules on it with an overall capacity of 1,840 Wp. Combined with a battery bank and predominantly DC appliances, the system was able to run completely independent from the grid.

The business had three income streams. The people who used the Internet paid an hourly fee, just like any other Internet café. On top of that, the centers provided computer classes. In addition, there was a large flat screen TV that served as a cinema (see Fig. 5.6). This was used for educational purposes and also at night to show soccer matches. The latter was actually important in terms of revenue because it generated additional income from the sale of snacks and drinks.

NICE started with two centers at two different locations (Fig. 5.7 shows the Banjul shop), and although these were operated by NICE Gambia Ltd, the idea was to roll out the concept as a franchise. The NICE concept attracted considerable attention from other African countries and we looked into introducing the franchise abroad. The management started fundraising for the international expansion in 2008, which looked promising. However, when the financial crisis hit, liquidity dried up very quickly and it was extremely difficult to attract funding for a venture like this.

However, NICE demonstrated it was possible to set up a viable business providing modern Internet and communication services in off-grid areas in Africa, 100% powered by renewable energy. Since then, appliances have become even more efficient, batteries more affordable, and solar modules much more powerful and cheaper.

Figure 5.6 The NICE cinema.

Figure 5.7 NICE shop, the Gambia.

5.6   Solar Energy and Mining

Africa is a continent with vast quantities of mineral wealth, much of it still untapped. Its mineral industry is the largest in the world. For many African countries, mineral exploration and production constitute significant parts of their economies and remain keys to economic growth, and in recent years, there has been a boom in mining operations on the continent. Gold mining is Africa's principal mining resource, and the continent ranks first or second in quantity of world reserves of bauxite, cobalt, industrial diamond, phosphate rock, platinum-group metals, vermiculite, and zirconium. Many other minerals are also present in quantity. In terms of production, Africa produces about half of the world's diamonds, one-fifth of the world's gold, one-sixth of the world's uranium, one-tenth of the world's bauxite (for aluminum), and 5% of the world's copper.

A major concern is that many African countries are highly and dangerously dependent on mineral exports. More than 90% of export earnings for Algeria, Equatorial Guinea, Libya, and Nigeria are due to exports of mineral fuels (coal and petroleum), and is 80% for mineral exports from Botswana, Congo, Gabon, Guinea, Sierra Leone, and Sudan. In addition, minerals and mineral fuels account for more than half of the export earnings of Mali, Mauritania, Mozambique, Namibia, and Zambia.

In the Middle East, which accounts for about 5% of the world's GDP, the hydrocarbon industry (petroleum, natural gas) continues to be the main driver of economic growth, directly through the wealth created by crude oil exports and indirectly through infrastructure projects in regional countries. The Middle East is a significant partner in the world's oil and gas trade, responsible for just under half of the world's crude oil exports and just over 40% of the world's liquefied natural gas exports. The region also holds about 10% of the world's petroleum refining capacity and accounts for about 13% of the world's refined petroleum exports. The region's abundant and low-cost supply of hydrocarbon fuels also gives it a competitive advantage for developing mineral industries to produce aluminum, cement, iron and steel, fertilizers, and petrochemicals. Exploration for other non-fuel minerals, focused mostly on chromium, copper, gold, lead, manganese, silver, and zinc, is also actively under way, most notably in Turkey as well as in Iran, Oman, Saudi Arabia, and Yemen.

Mining is energy-intensive—energy supply accounts for about 25% of a typical mine's operating costs—and growing costs of power and, most importantly, reliability of supply are the two biggest energy challenges facing the mining industry. Mines present a unique set of power requirements, especially for those operating off-grid: 24 hours of operation practically every day of the year with low tolerance for outages or shortfalls. An off-grid mine has no backup from a utility supplier, thus no ability to obtain reserve power, voltage support, or other grid services. The mine's power supply system must supply all of these services itself. Even when a mine is connected to the grid, that connection may be fragile and unreliable. To date these requirements have been met largely by generators using diesel fuel, which is expensive, especially when trucked to a mine over long distances. Combustion of diesel fuel also puts large quantities of carbon into the atmosphere, increasing business risks associated with environmental permitting and carbon tax regulations.

These requirements and challenges have provided an opportunity for the incorporation of renewable energy into mining operations, especially as the costs of solar and wind energy have decreased dramatically in recent years. This is despite the fact that mining executives are notoriously risk averse and reluctant to invest in assets that do not provide a payback within the life of a mine, typically 5–10 years. In addition, some mining executives are still unsure about renewables because of intermittency, the fact that the sun is not always shining and the wind is not always blowing. These concerns have been alleviated in recent years by the use of hybrid power supply systems that combine the outputs of different energy sources such as solar and diesel, so that when solar energy is available, diesel fuel requirements are reduced and when the sun is not available, a diesel (or natural gas or hydropower or wind) generator can meet the demand for energy. Increasing availability of low-cost battery and other storage systems is also enhancing the reliability of such systems and further reducing the need for diesel backup. In addition, a mine's renewable energy supply system, which has a typical useful life of more than 20 years, offers long-term energy cost stability, benefits local communities by creating jobs, can share power with local and distant communities, and can be left behind for local use after the mine has closed.

As a result of these potential benefits associated with the use of renewables in the mining sector, the reality that a hybrid power plant can be as reliable as or even more reliable than conventional plants, and as more and more experience with use of renewables in mining is gathered, a large increase in renewable energy applications in both on-grid and off-grid mines in the near future is expected.

One such example is Namibia, which currently imports more than 60% of its electrical energy from South Africa. A 2015 report (*REEE-powering Namibia*) examines the role that renewable energies could play in Namibia's energy future. Part of its analysis focuses on the mining industry, which is the most energy-intensive industrial activity in Namibia and accounts for almost one-third of Namibia's electricity demand. It identifies the multiple benefits that use of renewable energy in mining in Namibia could bring, a country blessed with renewable energy, including 350 days of sunshine per year: reliability of supply, the ability to lock-in long-term electricity prices, and reduced environmental impacts. Specific examples of solar and wind systems already in use in the mining industry in other parts of the world include the Los Pelambres Copper Mine in Chile (on-grid with 69.5 MW solar PV, 115 MW wind), the DeGrussa Copper/Gold Mine in Australia (diesel hybrid with 10.6 MW solar PV), the Gabriela Mistral Copper Mine in Chile (27.5 MW solar thermal), the Lac de Gras Diamond Mine in Canada (off-grid diesel hybrid with 9.2 MW wind), the Veladero Gold Mine in Argentina (diesel hybrid with 2 MW wind), and the Glencore Raglan Nickel Mine in Canada (off-grid diesel hybrid with 3 MW wind).

In the long term, analysts see a dramatic shift to renewables, particularly solar, as prices continue to fall, battery storage improves and becomes more affordable, and renewables become the dominant source of energy in the mining industry.

5.7 Tele-Medicine and Tele-Education

As discussed in Chapters 5.4 and 5.5, solar energy is being used to enable the widespread utilization of telecommunication and Internet technologies. While any electrical energy source can be used for this purpose, the availability of high solar insolation levels throughout the African continent and the Middle East, together with much reduced costs of solar panels, will establish solar energy in the future as a dominant source of electricity in these parts of the world. Two applications of these technologies critical to the growing economies of African and Middle East populations will be tele-medicine and tele-education. Access to these services is also an important pillar of political development.

Tele-medicine, also known as e-health, is the use of telecommunication and information technologies to provide clinical heath care to remote locations where these and other medical services are not regularly available. They permit rapid and convenient communication between patients and medical staff as well as accurate transmission of medical test results and imaging from one site to another. Other benefits include significant cost savings and support to disaster relief. While there were earlier forms of tele-medicine, today's tele-medicine is a product of late 20th- and early 21st-century advanced technologies.

Tele-education, also known as e-learning, is the use of audio, video, and computer technologies to educate students who may not be physically present at a school. It has been widely used to educate medical professionals and is now being applied to education of students at all levels in remote and non-remote locations. The convenience of Internet linkages to online information adds an important educational option even for students enrolled in regular schools. It provides the opportunity to train more people better and at lower cost.

It is important to note that tele-education is the modern version of distance education, which has been around for quite a long time. In 1728, Caleb Phillips placed an ad in the Boston Gazette seeking "...students who wanted to learn through weekly mailed lessons." In the 1840s, in the UK, Sir Isaac Pitman taught a shorthand course by mail, which included corrections on student submissions. Pitman's idea was so successful that, eventually, Sir Isaac Pittman colleges were established throughout the UK. In the United States, the idea was adopted in 1873 with the establishment of the first correspondence school.

The widespread use of computers and the Internet has made distance learning easier and faster, even to the point where virtual universities deliver full curricula online. The majority of U.S. public and private colleges now offer full academic programs online. Distance education comes in two broad categories: paced and self-paced delivery. Paced delivery is currently the most common mode in which students begin and end a course at the same time. Self-paced delivery programs allow for continuous enrollment, and the length of time to complete the course is set by the student and takes into account different learning styles.

Given the realities of life in many parts of Africa and the Middle East—limited to no access to education and medical care—solar-powered tele-education and tele-medicine offer the opportunity to significantly improve access to these services and provide a path to improved human welfare. This is particularly true for the people in Sub-Saharan Africa, who have the worst health, on average, in the world. According to the International Finance Corporation of the World Bank Group, the region has 11% of the world's population, carries 24% of the global disease burden, accounts for less than 1% of global health expenditures, and supports only 2% of the world's doctors and 3% of the world's health workers. Half of children's deaths under age 5 occur in Sub-Saharan Africa, and it has the world's highest maternal mortality rate. With its population expected to double in the next 10 years, it is estimated that $25–30 billion in new investments will be needed to meet the demand for medical services, and tele-medicine will play an important role.

Access is still the greatest challenge to health care delivery in Africa—less than half of Africans have access to modern health facilities. Other major challenges include counterfeit drugs, health care staffing shortages, and public corruption that diverts resources away from health care delivery. Tele-medicine addresses two of these challenges: access and worker shortages.

In addition to its use of telecommunications to educate health care workers, and its ability to eliminate the possible transmission of infectious diseases or parasites between patients and medical staff, the services delivered by tele-medicine can be broken into three main categories:

- **Store-and-forward** tele-medicine, which involves gathering medical data (e.g., medical images) onsite and then transmitting this data to a medical professional for assessment offline at a more convenient time. Dermatology, radiology, and pathology are medical specialties that are conducive to this form of tele-medicine.
- **Remote patient monitoring** allows medical professionals to monitor a patient remotely using various technological devices. Such monitoring is primarily used for managing chronic diseases such as heart disease, diabetes, and asthma and has application in home-based dialysis and joint management.
- **Real-time interactive** involves electronic consultations in real time between patients and health care providers. This method can be used for history reviews, physical examinations, psychiatric evaluations, and ophthalmology assessments.

These various possibilities have led to a variety of tele-medicine sub-specialties such as emergency tele-medicine, tele-nursing, tele-pharmacy, tele-rehabilitation, tele-trauma care, tele-cardiology, tele-psychiatry, tele-radiology, tele-pathology, tele-dermatology, tele-dentistry, tele-audiology, tele-pharmacology, and even tele-surgery using the emerging capabilities of robotic surgery.

Along with these many benefits, tele-medicine has its downsides: the cost of telecommunication and data management equipment and of technical training for health personnel, the potential for reduced human interaction between medical professionals and patients, an increased risk of error when medical services are delivered in the absence of a registered professional, the increased risk that protected health information may be compromised through electronic storage and transmission, the potential for poor quality of transmitted medical records, and the inability to start treatment immediately.

A useful example of what is possible can be gleaned from what has been done in recent years in India. EDUSAT (also known as GSAT-3), India's first solar-powered, dedicated educational satellite, was launched by the Indian Space Research Organisation (ISRO) in September 2004. It was placed in a geostationary orbit and its purpose was to meet the demand for an interactive satellite-based distance education system for the country. It went live in September 2005 and served until September 2010, when it was decommissioned, placed in a "graveyard orbit," and its functions transferred to other ISRO satellites. EDUSAT had, and its successors have, an extensive network of interactive and receive-only ground stations and have revolutionized classroom teaching. Similar applications are now taking place in other countries, and the use of telecommunications to facilitate health care is spreading globally as well but not yet as fast.

6

Financing: The Key to Africa and the Middle East's Solar Energy Future

6.1   Introduction

Section 6 addresses the holy grail of solar energy deployment in Africa, the Middle East, and elsewhere the availability of investment funds and the way financing is structured. It reviews the basics of such financing efforts by the United Nations, the World Bank Group, and other international organizations to provide investment funds and enable access to electrical services, the emerging concept of pay-as-you-go that is opening many new opportunities for solar deployment, and the use of large-scale auctions to reduce solar energy costs. The availability in recent years of affordable financing for energy projects in Africa is underwriting the continent's rapid economic development in the 21st century.

The Sun Is Rising in Africa and the Middle East: On the Road to a Solar Energy Future
Peter F. Varadi, Frank Wouters, and Allan R. Hoffman
Copyright © 2018 Pan Stanford Publishing Pte. Ltd.
ISBN 978-981-4774-89-5 (Paperback), 978-1-351-00732-0 (eBook)
www.panstanford.com

6.2   Solar for Energy Access in Africa

Richenda Van Leeuwen[79]

Africa today has a limitless amount of sunlight.
About 11 terawatts of electricity can be generated from the sunlight in Africa
alone, so we haven't even scratched the surface on that.[80]

—**Akinwumi Adesina, President, African Development**
Bank (AFDB), 2017

For many years, it was difficult to generate international interest to invest in renewable energy on the African continent. Despite its abundant natural resources and high levels of solar insolation for much of the continent that would appear to support renewable energy solutions, the prevailing viewpoint in the mid-2000s was that financial returns on conventional energy projects were far more robust. At that time, prohibitive system capital costs, including the high cost of solar panels, meant that there was an opportunity cost to investing in solar photovoltaic generation and most other renewable energy solutions. Exceptions were grid-tied hydropower, as well as small-scale solar photovoltaic installations such as in Morocco and thermal rooftop installations for water heating in several countries in North Africa.

Solar is a vanity. We need "real" electricity.

—Senior West African government official in conversation
with the author,[81] 2010

By 2016, the situation had changed enormously. According to the Bloomberg New Energy Finance Climatescope report,[82] clean energy investment across sub-Saharan Africa had nearly doubled in the 1-year period between 2014 and 2015, totaling US$ 5.2 billion. International Renewable Energy Agency (IRENA) data indicate that Africa's total solar PV installed capacity increased from around 500 MW in 2013 to 1,330 MW in 2014 and

---

[79]Richenda Van Leeuwen's biography is on page 242.
[80]https://www.businesslive.co.za/bd/companies/energy/2017-05-23-sas-renewable-energy-auction-system-the-best-in-africa/.
[81]Richenda Van Leeuwen.
[82]http://global-climatescope.org/en/region/africa/.

close to 2,500 MW by the end of 2016. Much of this increase was driven by South Africa, with roughly $4.1 billion of the total by the end of 2015. If measured in terms of megawatts, large grid-tied solar photovoltaic and solar thermal projects, from the 160 MW Noor I project in Morocco to large solar photovoltaic projects in South Africa, have in recent years attracted significant commercial debt and equity investment. This has been due in part to the following:

- economies of scale
- competitiveness of other markets leading developers to begin to look at African opportunities
- rapid drop in price of solar PV panels
- focus on improving regulatory and policy frameworks in Africa through the use of supportive mechanisms such as auctions and national feed-in tariffs (REFIT)

While these numbers and growth rates are impressive, in many other parts of the African continent, renewable energy is still very much emerging. Many announced projects remain in project evaluation phase. Clearly more work needs to be done in countries that have not adopted solar solutions at the rate of South Africa or Morocco.

6.2.1 "Below," "Beyond," and "Off" the Grid: Powering Energy Access

There is a second story on renewable energy in Africa beyond that of grid-tied generation capacity, one that has really emerged over the past 10 years. This has largely been a story of small-scale, off-grid and, increasingly, mini-grid solar PV solutions. Again, this has been driven in part by decreased prices for solar panels, combined with the availability of LED lights, and small-scale lithium ion batteries. It has also been driven by a number of companies and charities that were established over the past decade with a specific mission focus to develop and make small-scale solar solutions available to off-grid families. Early companies included such names as D. Light, Barefoot Power, Greenlight Planet, Stiftung Solar Energie and Solar Aid (the charitable arm of UK solar company, Solar Century). All these entities focused on delivery of small-scale off-grid products and solutions to off-grid households in developing countries in Africa and some in other parts of the developing world.

Urban and rural families have increasingly adopted a range of small-scale household and community level solar photovoltaic power solutions. These have ranged from "entry level" solar lighting products mentioned above, and more recently "pay-as-you-go" financing for purchase or rental of household and small business systems. This has resulted in a massive sea change in the way solar solutions are financed across Africa in the last 5 years, with newer companies utilizing mobile money solutions scaling rapidly in the provision of hundreds of thousands of small systems to off-grid consumers. Many have focused initially in East Africa and have included companies such as M-Kopa, BBOX, Mobisol, and others. Some of the product companies such as D.Light have now also moved to focus more on pay-as-you-go financing, since it allows consumers to rent, or rent to own systems, without a sizable initial down payment, required in more traditional bank financing of solar systems. In addition to these home systems, community-level solar and hybrid "mini-grids" have also seen an increased focus by policy makers and investors alike. Both of these system types are covered in more depth in Chapter 6.3. Solar-powered irrigation, including drip irrigation, is also increasing in some markets.

In this off-grid electrification sector, success is no longer measured in terms of megawatts installed. It is assessed in the life-enhancing services facilitated through enabling people to access electricity—often for the first time—which in turn support a range of development benefits related to education, health, and livelihoods. These include applications such as electricity to power modern health care service delivery, to light a home in the evening enabling a child to study, or the electrification of a livelihood that can lead to greater productivity and income. System sizes vary, from as little as 0.5 W to 100 W for home systems for off-grid home use, and to multiple, even tens of kilowatts or more for community use. These solutions have provided at a minimum basic lighting and energy services, allowing customers to transition away from dependence on dirty, dangerous, and relatively expensive kerosene and candles. These had remained the mainstay of household lighting in millions of homes across

Sub-Saharan Africa and in other parts of the developing world well into the 21st century. In 2010, tens of millions of people living without the grid were still using kerosene as their primary fuel for home lighting. According to the International Finance Corporation's Lighting Africa Initiative, African "Base of the Pyramid" (low income) households and small businesses spent more than $10.5 billion annually on lighting, with kerosene being the dominant fuel for that purpose.

6.2.2 Why Solar for Energy Access in Africa?

By 2016, across Africa, nearly 600 million people still did not have access to electricity, as reported by national electrification rates by the International Energy Agency in its 2017 World Energy Outlook. Nonetheless, this deficit was not evenly distributed across the continent. Rural electrification continues to lag urban electrification significantly, and regional variations exist across the continent. Central Africa still lags East Africa, which has seen significant progress since 2012, and which has become a focus area for many off-grid and mini-grid companies, with strong engagement in Kenya in particular, and increasingly Rwanda, Tanzania, and Uganda. Other countries, like the Democratic Republic of the Congo (DRC), where civil unrest remains, still have massive electrification deficits. Additionally, across Africa, many people whose homes and businesses are connected to the national grid, but for whom grid power remains unreliable due to a large number of power outages, have adopted solar power systems as a supplemental source of power.

6.2.3 Why Hasn't the Grid Been Extended across Africa?

A combination of factors has led to countries not extending the grid to cover their entire population. While some countries in Africa such as Algeria and Tunisia have achieved more or less universal grid access, extending the central grid to remote, low-population-density areas is relatively expensive. Poorer households generally have low demand profiles and therefore may not be considered a high priority for grid extension, particularly where a central utility is already struggling with financial viability issues. Successful grid extension programs require a financially and technically viable utility that can manage connection mechanisms for poorer households to connect to the grid, as well as payments, and address issues such as line losses from large-scale transmission. The IEA has estimated that, globally, an additional $30 billion per year on average will be required to provide universal access to electricity by 2030, far more than the $8 billion per year current level of investment. This includes projections around grid extension as likely to be the most suitable—and cost-effective—option for all urban areas and for around 30% of rural areas, with mini-grids and off-grid solutions completing the rest.

6.2.4 Global Catalysts: Renewed Attention at the UN and Beyond

Significant renewed attention to the issue of energy access has taken place in the international development community since 2010, with a strong push to recognize—and appropriately value—the close correlation between economic development, social benefits, and the availability of and access to electricity.

In 2011, former UN Secretary-General Ban Ki-moon, recognizing that at that time some 1.3 billion people around the world lacked access to electricity, launched his signature campaign, the "Sustainable Energy for All" initiative. The UN General Assembly declared 2012 to be the UN International Year of Sustainable Energy for All and 2014–2024 to be the Decade of Sustainable Energy for All. The aim of "Sustainable Energy for All" was—and remains—to support three goals: achieving universal access to modern energy services by 2030, increasing energy efficiency, and doubling the share of renewable energy in the global energy mix by 2030.

In 2015, UN member states adopted the Sustainable Development Goals, including Goal number 7, focused on achieving universal energy access by 2030. This was a significant win for the development community and for renewable energy advocates, many of whom believed that energy had been a "missing" component of the earlier UN Millennium Development Goals.

Energy is the golden thread that connects economic growth, social equity, and environmental sustainability[83]

—Former United Nations Secretary-General,
Ban Ki-moon, 2012

The UN initiative also drew increased attention to the need for a range of solutions not limited to the provision of grid electricity. As part of its contribution, the World Bank and partners developed a "multi-tier" framework for measuring energy access, focusing on the number of hours of electricity per day, and indicative types of power provision available in a given "tier" from Tier 0 (no access) to Tier 5 (highest level of access) (Fig. 6.1).

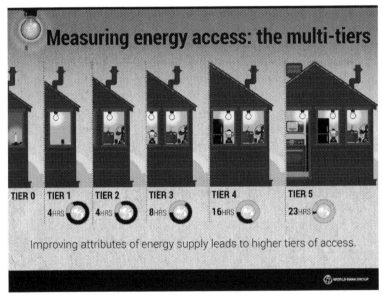

Figure 6.1 UN Sustainable Energy for All: Energy Access Multi-Tier Framework.[84]

Tier 0 is characterized as using candles or kerosene for lighting, with no appliance use. Tier 1 includes 4 hours of lighting, as represented by a small light and cell phone charging.

[83]https://www.un.org/press/en/2012/sgsm14242.doc.htm.
[84]Mikul Bhatia, Niki Angelou (2015). *Beyond Connections: Energy Access Redefined.* ESMAP Technical Report 008/15. World Bank, Washington, DC. © World Bank. https://openknowledge.worldbank.org/handle/10986/24368. License: CC BY 3.0 IGO.

The follow-on tiers include Tier 2, also with 4 hours, but adding small appliances such as a small solar LED TV or a fan, with these being powered by a home solar system. Tier 3 moves to 8 hours of electricity, including additional small appliances. Tier 4 also includes refrigeration in addition to lighting and small appliances, at 16 hours per day, which could possibly be powered by a larger micro-grid. Tier 5 essentially represents full energy access, with 23 hours per day of electricity. Tier 5 is assumed to mean having access to grid power, with enough reliability to power a full range of household lighting and appliances close to 24 hours per day.

The intention was never to define minimum levels for the provision of power. Instead, it was to recognize that for households starting from a very low base of using candles and kerosene for lighting (in some cases not even that), an initial transition to a small solar light represents an improvement in access, albeit not full access. The framework provides a methodology for measuring the availability of power (but not the quality) rather than focusing on specific appliances that are powered, since they will vary by consumer and context.

Global attention around energy access helped to inspire national initiatives, such as the US government's "Power Africa" initiative launched by US President Barack Obama in 2013. Power Africa's goals include providing 30,000 MW of new and cleaner power generation and 60 million new electricity connections. Its "Beyond the Grid" off-grid electrification sub-initiative specifically focuses on household solar and micro-grids, with the aim to add 25–30 million new connections (of the overall 60 million goal of Power Africa) by 2030. This initiative, recognizing that financing remains a key constraint in many markets, has taken a transaction-focused approach to energy access, seeking to address market barriers and financing issues within the context of an individual project or transaction. By 2016, Power Africa reported that its public and private partners had committed more than US $50 billion, including US $40 billion from private sector partners, to the initiative (see also Chapter 4.3).

Countries across Africa have embraced their own national targets for electrification and are focusing on different targets related to their SDG7 commitments, their Sustainable Energy for

All commitments, or their 2015 Paris Climate Accord "Nationally Determined Contributions" (NDCs), or some combination of all of them. Increased utilization of GPS planning tools also plays a role in a number of African countries, to help model "lowest"-cost options comprising both on- and off-grid solutions for planning purposes, although cultural factors and consumer preferences play a role in addition to cost and enabling environment.

The UN initiative was not the first development initiative to focus on bringing small-scale solar PV to sub-Saharan Africa. In addition to an earlier International Finance Corporation Photovoltaic Market Transformation Initiative (PVMTI), whose strongest success in Africa was, arguably, in Morocco, by 2008 the World Bank and the IFC had formed the "Lighting Africa" initiative. This sought to take the market well beyond North Africa and help to support bringing modern lighting by 2030 to off-grid African families. Lighting Africa has focused on the need to develop "a market" for affordable small-scale lighting and power solutions to be made available to low-income communities lacking grid power. One early aspect they focused on to do this was to support market studies in several countries in Africa to determine demand and provide needed data analytics to companies that were largely still at that point too small and early stage to be able to invest significant funds into market research. During this time, there was also a new focus on support for small-scale solar lighting products, entry-level products that were much smaller than the solar home systems, and typically used 1–3 W of solar power. A strong focus was taken on quality assurance (details are provided in Chapter 6.3) to help ensure the quality of the solutions being provided, given that a low-income customer would struggle to purchase a replacement product if their initial purchase was a fake brand, was defective, or broke soon.

6.2.5 Market Expansion

Apart from a number of pioneer companies, such as D.Light, Barefoot Power, and Greenlight Planet, many newer entrants have entered the sector over the past 10 years, including a plethora of others offering small-scale lighting products, small solar home systems, and larger systems for community applications. These were again enabled in part by dramatic cost decreases in the price of solar PV panels in recent years, the inclusion of LED lighting rather than incandescent bulbs or even compact fluorescent bulbs (CFLs), improvements in battery technology, and in software and in appliances. Applications for the use of solar for drip irrigation were explored through non-profits showing improvements in agricultural productivity,[85] and solar power water pumping began to be explored (see Chapter 5.2). Unfortunately, low-cost fake products have also increased in many markets, undermining sector quality and reputability.

By 2010, while the recession across much of the world decimated markets, in a sense it helped catalyze the market for solar PV in the developing world. In part driven by the European market for solar PV plummeting in 2009 due to the recession and abrupt changes in tariff regimes in some key markets such as Spain, solar panel manufacturers began expanding their reach into new parts of the world. Small companies that had previously struggled to acquire the panels that they needed for small-scale installations suddenly began to have more traction in the market, and panel makers began to look at markets they had previously ignored. This led to increasing availability of a range of sizes of panel suited to small-scale applications, and not only the standard 200 Wp sizes used in most grid-tied projects.

This chapter would not be complete without a mention of consumer household and small business appliances. The fast pace of innovation and the advent of super-efficient appliances, combined with smart software, have enabled even small amounts of power to provide significant benefits to a household at low cost. When combined with financing that allows for payment over time, these appliances become accessible to much larger market segments. Appliance makers in the past did not really cater to the off-grid market. The growth of such appliances as award-winning low-wattage DC (and AC) LED color TVs, now increasingly available in many African markets, as well as super-high efficiency fans, means that some of the services previously out of reach to poor homes are now becoming more financially accessible. Not all of these appliance makers today are large global brands like Samsung. One smaller German company manufacturing solar appliances is Fosera, which produces an

---

[85]http://self.org/sustainableenergyforall/.

array of solar products and systems. Among its product lines is an award-winning 12 V solar-powered LED television, which only uses 6.5 W of power, highly compatible with even a small DC solar PV system. The company sells its products in 11 African countries.

This trend of growth in off-grid and mini-grid systems is likely to continue and, if embraced by governments, may allow low-income households to adopt a better quality of lifestyle. This does not automatically necessitate a correspondingly high kW per capita usage, or even grid access, as was the case in OECD countries. Some off-grid solar companies will now include financing for appliances within their overall pricing structure for consumers, recognizing that what the customer wants is not the electricity as such, but the use of the appliance that is powered by it. This is an area that is likely to see significant further change in the coming years, particularly with the dawn of the "Internet of things" (IOT). This sees smart appliances that gather enormous amounts of data being interconnected to each other and the Internet. While there are potential vulnerabilities associated with this (such as system hacking), it can in principle help to tailor services more specifically to consumers in these markets over time.

6.2.6   Future Directions

While price decreases have led to massive increases in the number of people across Africa being able to adopt some type of solar energy access solution, the size of the residual challenge remains immense. Financing remains an issue, although unevenly in different parts of the market and the continent. Regulatory policy, including import tariffs, can also help to drive or hinder market development. Even in countries with well-developed regulatory structures, their existence alone does not mean that the market will develop well. Political engagement will remain important as will educating financial institutions on the range of opportunities and financing types needed by solar companies. This includes how this sector and individual projects can be successfully de-risked, and either funded commercially or through a well-tailored package of commercial and concessional financing— and, of course, ensuring that consumers can afford to access solutions that meet their needs and reach suitably high quality standards through appropriate quality controls as well as tailored consumer financing mechanisms, whether pay-as-you-go or through other approaches.

Above all, if the 2030 goal of achieving universal energy access is to be achieved across Africa or in other parts of the developing world, a clear and continuing goal should be to ensure consumers are not simply getting access to a solar-powered electricity solution but can utilize it successfully to support and maintain an improved quality of life.

6.3 Financing Solar in Africa and the Middle East

The investment in a solar energy system is usually made using a combination of equity and debt. Since banks are fundamentally risk averse, the debt equity ratio decreases with an increase in the risk environment. In very stable and developed markets, up to 90% of the total investment can be borrowed, whereas in conflict zones, often no credit is available and everything has to be paid with equity. In most cases, however, the debt equity ratio is between 50% and 80%. Other types of financing include leasing or lease–purchase schemes, which are often used in pay-as-you-go schemes.

Since most of the cost of ownership of a solar energy system lies in the upfront investment, the sources of capital and the way the financing is structured are important parameters in the overall cost of the service the solar system provides. Although falling renewable energy technology costs have significantly lowered the capital needed to invest in new systems, financing renewable energy projects is still difficult in many parts of the world. There are major differences and the cost and availability of financing depends on the region where the investment is taking place, as well as on the type of investment. For example, financing a large grid-connected solar PV system that supplies electricity to a government entity with a solid credit rating in one of the Gulf countries is relatively straightforward. The recent world record low tariffs for PV electricity in the United Arab Emirates and Saudi Arabia were partly due to historically low costs of borrowing in those markets. Furthermore, the currencies in those markets are pegged to the US dollar, so the tariffs can be used to repaying the investment, which is also typically in US dollar, without currency hedging, which typically adds to the cost. Such favorable circumstances cannot be found elsewhere in the Middle East or Africa, where a different environment for financing prevails. There are multiple and interesting possibilities of financing solar systems in Africa. This section aims to provide a snapshot of how solar is currently being financed in Africa.

6.3.1   Size Matters

In terms of overall investment, it should not come as a surprise that the bulk of money for solar in Africa is invested in large grid-connected solar plants, simply because they comprise the bulk of megawatts deployed. As reported by IRENA (see Section 8), at the end of 2016, some 2.5 GW of solar PV capacity has been installed in Africa, roughly a quarter of which is off-grid. So, 75%, or 1.9 GW, comprises grid-connected systems, the majority of which is in South Africa. According to Bloomberg's Climatescope,[86] which tracks most sizeable investments, from 2011 up to 2015, more than US\$ 24 billion was invested in renewable energy in sub-Saharan Africa; US\$ 11.3 billion, or 47% of that, was investments in solar energy.

As described in Chapter 6.4, South Africa has successfully introduced ever more economical solar PV through a well-planned system of consecutive auctions. However, it is important to note that even though South Africa's economy lies well ahead of the average economy in sub-Saharan Africa, the entire US\$ 7.3 billion that was lent to solar energy projects in 2012 and 2013 was provided by local South African banks. According to Citibank, for them to bid from abroad, including a currency hedge would have added 400 bips, or 4%, to any dollar amount provided by them or any other international bank, making them uncompetitive. Given the size of the South African economy and the liquidity in the banking sector, this was not a problem. However, local banks in other sub-Saharan African countries are, first, not very experienced in lending to renewable energy projects and, second, not as strong as the South African banks. Also, the investment climate in most countries and the requirement for long loan tenures make it difficult for developers of such projects to secure commercial finance. For this reason, international financial institutions such as the World Bank, the African Development Bank, the European Investment Bank, the Islamic Development Bank, and the Green Climate Fund provide funding to solar energy investments in Africa. Chapter 4.4 describes the World Bank's efforts in this respect. However, adding an international financial component

---

[86]http://global-climatescope.org/en/download/reports/regions/climatescope-2016-africa-en.pdf.

adds complexity to the process, which comes at a cost. The entire process of due diligence, risk management, negotiation, etc., is intrinsically expensive, especially in an international setting, leading to relatively high costs for such transactions. It then becomes difficult for smaller or medium-size projects to obtain such concessional financing. Many solar projects are relatively small. A 5 to 20 MW solar power plant, which in many sub-Saharan African countries would be sizeable, costs much less than US$ 50 million these days. This is not much for a project finance structure, making it difficult for most International institutions to engage, simply because their internal costs are too high for such investments.

Apart from South Africa, Morocco should be mentioned if large-scale solar investments in Africa are discussed. Although predominantly concentrating solar power (CSP), Morocco has successfully mobilized billions of US dollars in the Noor and Midelt solar complexes. Morocco's MASEN, which not only co-invests but also guarantees the power offtake on behalf of the Moroccan government, is a unique entity aimed to foster bankable investments in Morocco. Masen's structure and the size of the investments—Noor 1, 2, and 3 together amount to 510 MW—enabled access to a concessional finance package resulting in weighted average cost of capital (WACC) of slightly more than 4%, lower than anywhere else in Africa. This package was provided by the World Bank, the European Investment Bank, and others. As a result of this international support, CSP could come down in cost, recently leading to Dubai's 7.3 USc/kWh for a 700 MW CSP project, giving the CSP technology a new lease on life. These examples show that financial support for large-scale solar energy investments in Africa and elsewhere have served their purpose and brought down cost over time.

Important approaches to reducing transaction costs are to work toward standardization and pool investments. The main document in a grid-connected solar PV project is the power purchase agreement (PPA). IRENA and several other international organizations have started working on a standard PPA document, which aims to be instrumental in lowering the transaction costs of renewable energy investments. Equally, pooling smaller investments can increase the volume of the overall financing package to lift it over the minimum barrier for banks and investors.

Such mechanisms could also help securitize renewable energy assets for the purpose of trading in capital markets, like credit card debt or mortgages.

6.3.2   Risk

Risk or the perception thereof is one of the main topics determining the cost of capital for a renewable energy investment. It can best be described as a piece of art: The perception differs from person to person and depends on how you look at a subject. According to IRENA,[87] a perception of high risk constrains the development and financing of renewable energy projects. Project risk, perceived or real, can take multiple forms. These include political and regulatory risk; counterparty, grid, and transmission link risk; currency, liquidity, and refinancing risk; and resource risk. Public finance institutions such as the World Bank and the African Development Bank have a range of tools to support private investors with risk mitigation instruments. These include guarantees, currency hedging instruments, and liquidity reserve facilities. They help mobilize private capital while reducing the capital requirements of the public finance institution. However, according to IRENA, to increase the use of risk mitigation instruments, these organizations need to simplify procedures, reduce transaction costs, set internal incentives, and further expand their toolbox with instruments specifically targeting renewable energy project needs. More needs to be done to make financing available for renewable energy investments throughout Africa.

6.3.3   Financing Off-Grid

At the end of 2016, Algeria, Egypt, Ethiopia, Kenya, Morocco, South Africa, and Uganda all had more than 20 MW of off-grid capacity installed. Some of that capacity is found in sizeable mini-grids (Egypt, Morocco, Algeria), while in South Africa off-grid may not be truly off-grid (standalone) either, but backup systems and panels installed to reduce electricity bills.

---

[87]https://www.irena.org/DocumentDownloads/Publications/IRENA_Risk_Mitigation_and_Structured_Finance_2016.pdf.

Chapter 6.4 describes the main business models seen in the off-grid solar market in Africa. An important growth sector is the pay-as-you-go (PAYGO) solar concept. Although PAYGO companies constitute just about one-fifth of all companies in the off-grid solar sector, the companies that transitioned from cash-sales to PAYGO saw a substantial increase in customer uptake, providing evidence that this is a game-changing addition to the business model. According to Bloomberg New Energy Finance, there are typically four stages of financing PAYGO companies: the seed stage, the early stage, the expansion stage, and the scale-up stage. In the seed stage, companies look for funds up to $1 million, usually from angel investors or grants. This is mostly equity. In the early stage, companies such as PAWAME typically seek amounts ranging $3–5 million, also mostly equity. In the expansion phase, firms such as M-Kopa, Off-Grid Electric, D.Light, BBOXX, Nova Lumos, Fenix International, and Mobisol have all announced at least one transaction larger than $10 million, in a combination of equity and debt. Several companies have now reached the scale-up phase and they are looking for amounts ranging $50–100 million, which would be debt. Given the limited balance sheets of these companies, an important element would be some form of securitization of assets. Given that this asset class is new, and much of the value is dispersed over thousands of households in rural areas in developing countries, it probably requires some time before lenders understand this and get comfortable with the risk profile. Nonetheless, this mechanism can potentially develop very fast. Residential solar securitization in the United States rose from $53 million to $803 million in just 2 years.

There is anecdotal evidence, however, that the pool of capital provided by venture capital funds is drying up. The business model is such that a large amount of money is required upfront. A substantial investment in the back office, sales, and installation staff and, of course, all the solar systems is required. Although the customers have to make a down payment, most of the investment is recouped through small payments over a longer period. This means that it will take several years before such company returns a solid profit, especially since most of them are growing aggressively. Since many companies started around the same time, the investor community would first like to see proof that the business model works before investing more money.

Traditional donors seem to have a bias toward larger cash-generating projects such as power plants with a power purchase agreement. However, a mini-grid in a rural area or a PAYGO Company is a business, rather than a project. This makes it more complicated to tap into traditional sources of concessional finance. Several international financing institutions have recognized this problem and either established dedicated funds or invested in existing funds. One example is KawiSafi Ventures, anchored by the Green Climate Fund, which focuses on the off-grid energy sector in East Africa to provide universal access to energy to people located.

Solar energy is one of the most universal energy sources, capable of providing modern energy to even the farthest corners of Africa and the Middle East—corners where the grid struggles to reach because of the low population density. Solar PV is one of the most competitive energy sources nowadays, and there beyond the foreseeable grid connection and at the bottom of the economic pyramid are proven and viable business models for the entire range of applications, from the smallest solar lantern to megawatt-scale power plants. However, many companies and end users are often struggling to access capital to finance their systems, despite the undisputed benefits for the economy and development of remote communities. What is required is a concerted and focused effort by the financial community to provide innovative financing solutions that match the innovation of the solar business community.

6.4   Pay-As-You-Go and Community Solar

6.4.1   Where the Grid Doesn't Reach

More than 620 million people in sub-Saharan Africa are not connected to the grid, and this number is rising despite spectacular economic growth in many countries. The number of people living off the grid has risen by more than 114 million on the continent since 2000, with several million more joining every year. Although many governments are working on expanding their electricity grids, many people live in sparsely populated areas far away from main transmission corridors. For such areas, extending the grid is a very expensive option, costing between $500 and $1,000 per connection, and the expected demand for electricity will be limited for years to come. Dedicated off-grid solutions are in many cases a more economical and appropriate solution. The standard solution so far has been a diesel generator. The recent cost competitiveness of renewable energy, especially solar PV, has made business models based on renewables much more compelling. Renewables don't need fuel and the lack of moving parts in the case of solar PV, reduces the need for maintenance. Several solutions using modern renewables to provide off-grid electricity have been around for some time. They include solar products such as solar lanterns, solar home systems, and mini-grids, but also dedicated solutions to power a business.

6.4.2   Solar Products

People that do not have access to grid electricity will meet their energy needs by other, often expensive, means. According to a report released early 2016 by the World Bank Group and Bloomberg New Energy Finance, in collaboration with the Global Off-Grid Lighting Association,[88] globally 240 million mobile phone subscribers live off-grid. These phones are often charged by small businesses costing on average $0.20 per charge. Expressed in electricity, this equates to an astounding $30–50 per

---

[88]https://www.lightingafrica.org/launch-of-2016-world-bank-group-bloomberg-off-grid-solar-market-trends-report/.

kWh. For a proper cost assessment, one would have to include the travel time to the charging point, which can be substantial. This shows the enormous potential for cost-effective off-grid electricity solutions, including solar. Figure 6.2 shows that people spend $2.4 billion on mobile phone charging in Africa each year. But they also spend over $14 billion on kerosene, batteries, and candles for lighting, an amount almost as high as the GDP of a country like Zambia. The annual expenditure bill for household lighting varies from country to country but ranges $100–140/year. Kerosene lanterns, a century-old technology, not only are costly but also are fire hazards. The wicks smoke, the glass cracks, and the light provided is often too weak to read by. The World Health Organization says the fine particles in kerosene fumes cause chronic pulmonary disease, affecting women and children.

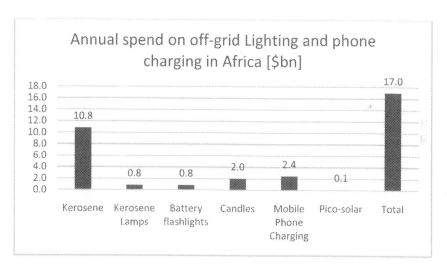

Figure 6.2 Estimated annual spend on off-grid lighting and phone charging in Africa (2014, billion dollars; source: footnote [89])·

However, in recent years, with solar panels becoming more and more affordable and LED lights becoming more and more practical, so-called solar lanterns have been selling in the millions.

---

[89]Fatih Birol, et al., IEA (2014). *Africa Energy Outlook: A Focus on Energy Prospects in Sub-Saharan Africa.*

Solar lanterns consist of an electric lamp (usually an LED) with a rechargeable battery that is charged with a small solar panel. If the solar panel is smaller than 10 W, the system is called a pico-solar system.

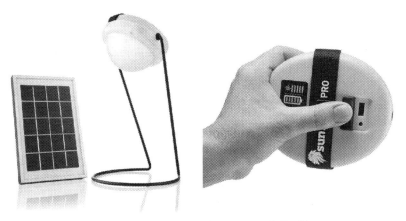

Figure 6.3 A solar lantern with phone-charging capability.[90]

Figure 6.3 shows a solar lantern with phone-charging capability. Even the most basic solar lamps outperform kerosene lanterns in terms of ease of use and quality of light. A typical solar lantern takes 8 to 10 hours to charge and then provides 4 or 5 hours of light from high-efficiency white LEDs. The number of times solar lamps can be charged before their rechargeable batteries wear out has improved enormously in recent years, along with their ability to cope with dust, water, and being dropped. A simple solar lantern can cost as little as $5, with higher-quality models, for example, with phone-charging capability, costing up to $50. Although this is not cheap for a poor rural household, the savings on kerosene and mobile phone charging provide for compelling economics. A "typical" solar lamp costing $13 has a payback time of just a few months in most sub-Saharan African countries. Assuming a life of 2 to 3 years for a product from a reputable manufacturer, customers can enjoy at least 1 year of free lighting until they have to make another investment.

---

[90]http://sunnymoney.org/index.php/solarlights/.

Until the end of 2015, more than 20 million units have been sold in Africa, with half of those supplied by reputable companies with a quality product, the other half comprising lower-cost generic products. In the second half of 2016 alone, an additional 1.87 million solar products were sold in Sub-Saharan Africa.[91] It is no surprise that quality products, requiring a price premium, have been copied by manufacturers of generic products, with lower-quality components and a consequential shorter life or inferior service.

Lighting Global[92] is the World Bank Group's platform supporting sustainable growth of the off-grid solar market. Through Lighting Global, the International Finance Corporation (IFC) and the World Bank work with the Global Off-Grid Lighting Association (GOGLA), manufacturers, distributors, and other development partners and end users to develop the off-grid lighting market. An important aspect is the quality of the products. Lighting Global provides a quality assurance methodology with the following key aspects:

- **Truth in Advertising:** Advertising and marketing materials accurately reflect the tested product performance.
- **Durability:** The product is appropriately protected from water exposure and physical ingress and survives being dropped.
- **System Quality:** The product passes a visual wiring and assembly inspection.
- **Lumen Maintenance:** The product maintains at least 85% of initial light output after 2,000 hours of operation.
- **Warranty:** A 1-year (or longer) retail warranty is available.

Quality is important, since a survey done by SolarAid in Tanzania,[93] where a lot of low-quality products are being sold, showed customer satisfaction slipping from 97% for quality-verified brands to 60% for other brands due to lower product quality.

---

[91]*Global Off-Grid Solar Market Report: Semi-Annual Sales and Impact Data*, July–December 2016, Public Report, GOGLA, Lighting Global, IFC, Berenschot.
[92]https://www.lightingglobal.org/
[93]SolarAid (June 2015). *Research Findings: Baseline and Follow-Up Market Research in Kenya, Tanzania and Zambia.*

6.4.3 Solar Home Systems

One step up from pico-solar are solar home systems. In 1992 Dr. Anil Cabraal of the World Bank ASTAE division came up with the idea (Chapter 4.4), which was named "Solar Home System" (SHS). These systems were believed to be a potential solution for the 1.6 billion people who do not have grid electricity. Solar home systems are individual household systems consisting of a solar panel, a battery, associated control circuitry, and one or more domestic appliances.

Compared with conventional energy available in rural areas, such as kerosene and disposable batteries, the cost of a solar home system must be paid up front. This typically requires some sort of financing, since most people in rural areas of developing countries cannot afford such a large amount. Although most developing countries have systems of rural credit supply, most traditional loans are given toward productive uses, e.g., to buy seeds for the next year's crop, assuring the bank there will be a future income to pay back the loan. Although clean household energy for light, mobile phone charging, and radio/TV supports development in the long run, it is difficult for a bank to identify additional income in the time they require to recoup the money they lend to the rural customer. So, solar home systems have traditionally been supported by development finance institutions (DFIs) such as the World Bank, the African Development Bank, and others, which supported local banks that had the infrastructure in place to provide such loans in rural areas. Also, 20 years ago, one needed big solar panels and heavy car batteries to power inefficient appliances for an individual household. However, now, with LED lights and efficient DC appliances, and low-cost reliable storage based on Li-ion batteries, which are much lighter and last longer than lead acid batteries, the overall energy demand is reduced. So a much smaller panel is required. And since the price of solar has fallen dramatically, the overall cost of the system is much more affordable now than 10–20 years ago.

6.4.4 M-Kopa

It wasn't until another innovation became available—mobile money—that we saw a real breakthrough in the market uptake of solar home systems in East Africa. M-Kopa[94] is the company that pioneered the application of mobile money in combination with solar systems, first in Kenya and now in more and more countries. M-Kopa basically provides a financing solution so that people can afford solar PV.

In 2006, Chad Larson and Jesse Moore were doing an MBA at Oxford University, when they met Nick Hughes. Nick presented M-Pesa, the first successful mobile money system as pioneered by him, when he was working for Vodafone in Kenya. The success of M-Pesa[95] intrigued them and they kept in touch with Nick in the years following his presentation. In the years between 2007 and 2010, they all had other jobs. In 2009, however, both Jesse and Nick left their jobs to explore business opportunities based on mobile banking. Both of them had a background in telephony and they were happy that Chad joined them in 2010, since he was a banker. In 2010, they successfully applied for grant money to pilot a few businesses based on mobile phones in Africa, among others from the Shell Foundation. They introduced a mobile savings account, they started a company providing health services by linking into a network of medical doctors, and they piloted a business providing solar home systems that incorporated a switch that can be operated remotely using a SIM card in a modem. Of the three ventures, the solar venture was by far the most successful. They installed 300 pilot systems and 95% of the people actually paid back their dues because of that switch. The additional response from neighbors, etc., was overwhelming. They had so many requests for additional systems that they realized that this was a winning formula.

Their system works as follows: The client can charge credit via the modem using his cell phone. If the credit used to pay for the solar system is depleted, the system automatically switches off. The entire system is automated, customers get an SMS warning them of low credit, and they will be warned again when the system is remotely switched off. The credit used is the normal mobile phone credit that can be bought at roadside stalls throughout the country.

---

[94]http://www.m-kopa.com/.
[95]https://www.worldremit.com/en/m-pesa-mobile-wallets?gclid=CO3Tu8qdxs0CF UFehgodNFUMkA&ef_id=Ue2rSQAAAY8sp3yU:20160626173650:s.

At present, people spend considerable amounts of money on kerosene and disposable batteries. M-Kopa targets rural customers who typically have one or two phones in the household. Most households spend half a dollar each day on kerosene and batteries, and M-Kopa charges that amount for a much better solar electricity service. With new technology, solar can now easily replace kerosene and batteries and supply enough energy for a DC radio, a flashlight, and even a TV. The technology as marketed by M-Kopa provides the experience of having grid electricity. The system is supplied with cabling and light switches, so when people enter their house, they can switch on the light immediately. A 20 W solar panel provides enough electricity to charge a lithium-ion battery, with a 5-year life, to watch 4 hours of TV on a 15° DC-powered TV set. In the future, M-Kopa is thinking about a small, highly efficient fridge. In terms of people's demand for electricity services, they want light, TV, phone charging, a fan, and possibly a fridge. All of these services can be competitively supplied with solar, which is more affordable then a grid connection. For such a grid connection in Kenya, a potential customer has to pay a deposit of US$ 400 to the grid company, and in addition there is a substantial tariff, consisting of a fixed and flexible rate.

After the success of the pilot project, the M-Kopa founders knew that to roll out a proper business, they needed additional money. They started a fundraising effort and managed to attract Gray Ghost Ventures out of Atlanta in the United States, which invested $1.5 million in the first round of financing, together with a few other institutional investors. This enabled them to design a back-office that could handle hundreds of thousands of micro transactions every day without a glitch. They also further refined the solar system and its components. Jesse and Chad moved to Kenya in 2011 and started hiring staff. As of May 2017, they employ 1000 full-time staff and 1,500 sales agents in East Africa. The company has more than 500,000 customers in Kenya, Uganda and Tanzania and adds 500 customers each day. After Kenya, they selected Uganda and Tanzania as next markets because the countries all have approximately 40 million people, low electrification rates, and most importantly, have well-introduced mobile money systems.

In Kenya, for example, people even buy roadside snacks and pay for them using mobile money. It is expected that in the next 10 years, many more countries will introduce mobile money systems. Countries such as Nigeria and Bangladesh have introduced mobile money and its adoption rate is growing fast, so they will be future markets for M-Kopa or other companies using mobile money.

There are an estimated 20 companies like M-Kopa, active in the pay-as-you-go or PAYGO segment, which is just about one-fifth of all companies in the off-grid solar sector. However, the companies that transitioned from cash-sales to PAYGO saw a substantial increase in customer uptake, providing evidence that this is a game-changing addition to the business model. According to Bloomberg New Energy Finance,[96] substantial new capital was injected in PAYGO companies in 2016, with more than $223 million of investment funds announced. This puts the sector well above the $158 million injected in 2015. Off Grid Electric, BBOXX, Nova Lumos, and Mobisol all raised individual rounds of at least $18 million.

---

[96]https://data.bloomberglp.com/bnef/sites/14/2017/01/BNEF-2017-01-05-Q1-2017-Off-grid-and-Mini-grid-Market-Outlook.pdf.

6.5   Large-Scale Auctions

6.5.1   Introduction

Governments have used several mechanisms to stimulate the deployment of renewable energy. European countries, most notably Germany, championed the use of feed-in tariffs. Feed-in tariffs provide renewable energy generators with a fixed price for the energy that they produce. These tariffs, which until recently were a fixture of Germany's Renewable Energy Act, were instrumental in the German "Energiewende" or energy transition and helped boost renewable energy's share in the national electricity production mix from less than 4% in 1990 to more than 30% today. Spain, another country that implemented feed-in tariffs, has steadily increased its renewable coverage of electricity demand from 18.4% in 2006 to 37.4% in 2015. A feed-in tariff system is relatively simple, provides certainty for the investor, and is characterized by low transaction costs. However, getting the tariff right is not easy and for a while Spain's tariffs were too generous, leading to windfall profits for developers. The Spanish government was unable to respond quickly enough to rapidly falling prices for most notably solar technology and consequently felt forced to introduce retroactive cuts to their feed-in tariff. Following Spain's example, more countries introduced similar retroactive cuts, which caused substantial confusion in the European market. Many investors started legal proceedings against European governments, and these court cases have still not been resolved. Although highly successful in bringing substantial renewable energy capacity online quickly, in recent years, more and more countries have replaced or complemented feed-in tariff systems with more market-oriented support schemes such as renewable energy auctions.

Structured price-based competitive procurement (auctions) of utility-scale solar PV has been a major factor in driving down the cost of solar PV in the past couple of years. There has been a substantial increase in the number of countries that carried out renewable energy auctions, from 7 countries in 2005 to 60 countries in 2015.

There are various types of auctions: sealed-bid auction, descending clock auction, and hybrid auction.

6.5.2 Sealed-Bid Auction

In a sealed-bid auction, bidders simultaneously submit a technical bid and a financial bid. The technical bids are checked whether they meet the minimum technical requirements, and the financial bids of the qualified technical bids are then opened and ranked. Projects are awarded until the sum of the quantities that they offer covers the volume auctioned. If it is a single project auction such as the 160 MW CSP Ouarzazate I project in Morocco (2011), the lowest bidder wins. The alternative is a pay-as-bid auction, which results in the allocation of multiple units of the same product with different prices. This can be to more than one project developer (e.g., the 3,725 MW auction of South Africa in 2010).

Often these auctions are carried out in two stages: a pre-qualification stage, resulting in a shortlist, and an evaluation stage.

6.5.3 Descending Clock Auctions

In a descending clock auction, the auctioneer starts with a high price and progressively lowers the price until the quantity offered matches the quantity to be procured. It differs from the sealed-bid auction in that it uses multi-round bids. Participants know each other's bids and adapt their price and quantities accordingly in subsequent rounds, which allows for a strong and fast price discovery, making it very efficient. Transparency also makes it less prone to collusion or corruption.

6.5.4 Hybrid Auctions

Some countries have introduced hybrid auctions, consisting of a first phase with a descending clock auction, allowing them to need supply with a certain margin and find a price, which can be used as a ceiling in a next phase, which is a sealed-bid auction, to meet the actual demand with the lowest price. Brazil has carried out such hybrid auctions.

Renewable energy auctions have become the instrument of choice for governments for the following reasons:

- Multi-project auctions typically have lower transaction costs than individually negotiated projects.
- The transparency associated with publicly run auctions will attract international investors.
- Auctions are quicker and less resource-intensive for the buyer.
- Auctions have been proven to yield lower pricing, especially in consecutive rounds.
- A bundled approach creates an opportunity to provide stapled finance with better financing terms, reducing the overall cost to the buyer.
- A structured process with clear oversight increases trust, reduces risk premiums, and leads to lower costs.

Countries that have achieved tariffs well below $0.10/kWh through such an approach until early 2017 include Mexico ($0.035/kWh), Jordan ($0.06/kWh), South Africa ($0.073/kWh), Peru ($0.05/kWh), Zambia ($0.06/kWh), the UAE—Dubai/Abu Dhabi ($0.03/kWh/$0.024/kWh), and Jamaica ($0.085/kWh). Several other countries are now considering auctions to procure cost-effective solar PV capacity.

Apart from the low cost, another main feature of PV is that it can be built extremely fast. The 75 MW Kalkbult solar PV power plant in South Africa was built in 8 months only, which compares extremely favorably with conventional power. A similar experience was made in Uganda, with KfW's GETFiT program. A 5 MWp PV system was built in 6 months, much faster than the 2–3 years required for hydropower projects developed under the same scheme. The low environmental and social impact and modular nature of the technology contribute to this fast deployment.

6.5.5 South Africa

In 2009, the South African government began exploring feed-in tariffs (FiTs) for renewable energy, but these were later rejected in favor of competitive tenders.[97] The resulting program, now known as the Renewable Energy Independent Power Producer Procurement Program (REIPPPP),[98] has successfully channeled substantial private sector expertise and investment into grid-connected renewable energy, including solar PV, in South Africa at competitive prices.

The following provides an overview over the first bidding rounds.

In August 2011, an initial Request for Proposals (RFP) was issued, and a compulsory bidder's conference was held with over 300 organizations attending. By November 2011, 53 bids for 2,128 MW of power generating capacity were received. Ultimately, 28 preferred bidders were selected offering 1,416 MW for a total investment of close to $6 billion. Major contractual agreements were signed on November 5, 2012, with most projects reaching full financial close shortly thereafter. The first project came on line in November 2013. A second round of bidding was announced in November 2011. The total amount of power to be acquired was reduced, and other changes were made to tighten the procurement process and increase competition. Seventy-nine bids for 3,233 MW were received in March 2012, and 19 bids were ultimately selected. Prices were more competitive, and bidders also offered better local content terms. Implementation, power purchase, and direct agreements were signed for all 19 projects in May 2013. A third round of bidding commenced in May 2013, and again, the total capacity offered was restricted. In August 2013, 93 bids were received totaling 6,023 MW. Seventeen preferred bidders were notified in October 2013 totaling 1,456 MW. Prices fell further in round three. Local content again increased, and financial closure was expected in July 2014. A fourth round of bidding was commenced in August 2014 and concluded with the award of contracts for 1,121 MW renewable energy capacity to six preferred bidders.

---

[97]Anton Eberhard, et al. (May 2014). *South Africa's Renewable Energy IPP Procurement Program: Success Factors and Lessons.* (http://www.gsb.uct.ac.za/files/ppiafreport.pdf).

[98]http://www.gsb.uct.ac.za/files/ppiafreport.pdf.

Figure 6.4 shows the tariff learning curve achieved over four consecutive rounds of competitive procurement in South Africa.

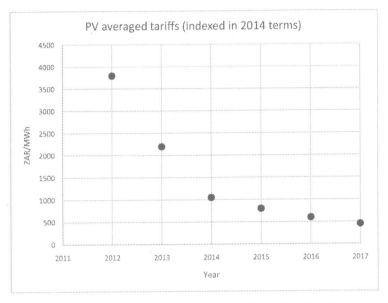

Figure 6.4 Development of electricity tariff in South Africa over consecutive competitive procurement rounds (ZAR = South African Rand (1 USD = ZAR 13.6)).

Similar tariff learning curves have been observed in many other countries in the world, making a strong case for a structured procurement approach with several previously announced auction rounds.

6.5.6 IFC's Scaling Solar

During his 2-year tenure as deputy director-general of IRENA, one of the authors (Frank Wouters) frequently came across the African renewable energy developer conundrum. When interacting with ministers, they often mentioned the lack of finance for projects in their countries. On the other hand, international financial institutions complained about the lack of bankable projects. An analysis of this paradox shows that typically the problem lies with the (local) developers of these projects, who often lack critical skills. It is not so much the lack of project opportunities, but rather the lack of capacity to turn such opportunities into bankable projects. Compared with other technologies, solar PV is relatively straightforward and not as complex as wind, geothermal, or hydropower projects.

With these things in mind, a group of people at the World Bank Group's private sector arm, the International Finance Corporation (IFC), started the development of a standardized approach to accelerate the deployment of utility-scale solar PV projects throughout Africa, which they called Scaling Solar. It began to take shape in 2013, when IFC chief investment officer Jamie Fergusson, principal investment officer Yasser Charafi and senior investment officer Dan Croft, who was then working in the IFC's advisory division, realized they were all working on a similar concept but coming at it from different angles. Croft, Fergusson, and Charafi believed that solar projects were fundamentally straightforward. They had analyzed how typical auctions were structured and had a close look at the South African experience. They realized that other sub-Saharan African countries require a lot more support than South Africa, and they developed a set of documents and procedures that could be replicated easily. South Africa has a strong banking sector and when the government designed their procurement process for renewable energy projects, local banks were involved. Consequently, the documents could be presented as non-negotiable, because they were bankable from the outset, saving a lot of time that would otherwise be spent negotiating details in the contracts with different developers. It was this element the IFC team wanted to replicate. The IFC constructed a package involving other divisions of the World Bank Group, including the Multilateral Investment Guarantee Agency (MIGA). The package has been designed to provide security to governments, bidders, and financiers. Before the auction starts, the Scaling Solar team will do all due diligence using donor funds rather than expecting the bidders to do their own, saving costs for the developer. In addition, the World Bank will provide a partial risk guarantee, which means governments need not provide a sovereign risk guarantee to satisfy financiers, although the template does include a government support agreement with direct government undertakings. The result is a process that substantially reduces the risk for the developer and the banks and that can be carried out very fast, which should lead to lower-cost solar electricity.

6.5.7 Zambia

A recent example of a successful Scaling Solar PV auction is Zambia.

Zambia is a landlocked country north of Zimbabwe. Of its total installed electricity generation capacity of 2,347 MW, hydropower is the most important energy source in the country with 2,259 MW (96%), followed by diesel contributing about 4% to the national energy supply. However, the low rainfall in 2015 has resulted in a national power generation deficit of about 560 MW. Scheduled power outages were having a negative impact on homes and businesses, and in 2015 Zambian President Edgar Lungu directed Zambia's Industrial Development Corporation (IDC) to develop at least 600 MW of solar power in the shortest possible time to address the power crisis. The IDC is an investment company wholly owned by the Zambian government, incorporated in early 2014, whose mandate is to catalyze Zambia's industrialization capacity to promote job creation and domestic wealth formation across key economic sectors. The IDC plays its role by serving as co-investor alongside private sector investors. The IFC has worked with the IDC on the development of two 50 MW solar PV independent power projects in Zambia, following the IFC's Scaling Solar initiative.

The two locations were pre-developed, meaning that the grid connection studies were done, the solar resource was assessed, and a complete finance package was provided by the IFC in the form of stapled finance. The term stapled finance describes a pre-arranged financing package that is "stapled" to the documentation and offered to the developers to use (they are not obliged). They then carried out a two-stage auction, with a pretty strict qualification phase. As a result, a select number of credible developers could focus fully on providing their best price for the solar electricity. The results, which came out in May 2016, exceeded everybody's expectations. The winning bids were for just 6.02 cents per kWh and 7.84 cents per kWh—the lowest prices for solar power to date in Africa, and among the lowest recorded anywhere in the world. Because the tariff is fixed for 25 years and will not rise with inflation, it represents about 4.7 cents per kWh over the life of the project—on par with recent auctions in Peru and Mexico and lower than the lowest tariff achieved for solar in South Africa so far. The two new solar power plants will increase the country's available generating capacity by 5% and will also help restore water levels in its Kariba and Kafue Gorge dams.

On February 21, 2017, the IDC signed an agreement with the IFC to develop up to 500 MW of clean, renewable energy through two to four additional projects.

In addition to Zambia, the IFC is working with the government of Senegal to develop 200 MW, with Madagascar to develop 30–40 MW, and with the government of Ethiopia to develop up to 500 MW of solar PV.

6.5.8 Epilogue

Auctions for large-scale solar have proven to be an effective tool for governments to increase solar generation capacity quickly and transparently. Auctions are among the best mechanisms to find the current price of a certain technology, and a set of well-structured consecutive auctions is instrumental in building a local knowledge base and reducing the cost over time. As has been proven by the case of Zambia, the IFC's Scaling Solar approach has enabled finding a tariff in sub-Saharan Africa that is comparable with global benchmarks.

Recent results have shown that auctions can be structured in specific ways to achieve desired outcomes. In general, achieving the lowest possible tariff is an important driver behind auction design, and an analysis of the lowest achieved prices, most notably the recent records achieved in Dubai and Saudi Arabia, yields a number of important elements that have contributed to such price levels. Carrying out a substantial amount of pre-development work reduces development risk and helps developers focus their efforts on building a project most cost-effectively. This includes designating and potentially preparing a site. In Dubai, the land was offered for free. Furthermore, one should consider carrying out a bankable resource assessment, geotechnical and seismic studies, offering grid-connection, etc. The more such elements are taken away from the developer, the risk profile is reduced and the lower the tariff will be.

7

Local Value Creation

7.1 Local Value Creation: Analysis

According to IRENA, until early 2017, close to 10 million people were employed in the renewable energy sector globally.[99] Unfortunately, Africa as a continent employs only 62,000 people in the sector, almost half of whom are in South Africa and one-fourth in Northern Africa. Africa, the world's youngest region, continues to be confronted with high levels of unemployment, vulnerable employment, and working poverty with little signs of potential recovery in 2017, according to a new study of the International Labour Organization (ILO)[100]. Despite Africa's impressive economic growth in the past decade, many (young) people still live in poverty, even if they are working. The working poverty rate among youth in Sub-Saharan Africa is nearly 70% in 2016, translating to 64.4 million working youth in that region living in extreme or moderate poverty, surviving on less than $3.10 per day. The region continues to report the highest youth working poverty rates globally. The number of poor working youth has increased by as much as 80% over the past 25 years. Northern Africa paints a slightly better picture, where a quarter of working youth are estimated to be living in extreme or moderate poverty, which represents a significant improvement since 1991, when almost half of all employed youth were living in poverty. It is therefore no surprise that African governments are looking toward renewable energy policies that maximize employment.

One of the policy instruments most widely deployed are Local Content Requirements (LCRs), either as a condition to receive public financial support or as part of eligibility requirements in public tenders. In that sense, they are often used in conjunction with (expensive) public financial support.

---

[99]https://www.irena.org/DocumentDownloads/Publications/IRENA_RE_Jobs_Annual_Review_2017.pdf.

[100] ILO, *World Employment and Social Outlook 2016: Trends for Youth*, www.ilo.org.

The Sun Is Rising in Africa and the Middle East: On the Road to a Solar Energy Future
Peter F. Varadi, Frank Wouters, and Allan R. Hoffman
Copyright © 2018 Pan Stanford Publishing Pte. Ltd.
ISBN 978-981-4774-89-5 (Paperback), 978-1-351-00732-0 (eBook)
www.panstanford.com

7.1.1 Local Content Requirements

Local content requirements dictate a certain percentage of the value of a renewable energy investment in a country to be sourced from domestic providers. A study carried out by the ICTSD and the Global Green Growth Institute[101] found four main drivers for LCRs:

- to augment public support for renewables
- to protect infant industries until they can compete on the international stage
- to create "green" jobs
- to secure environmental benefits through greater competition between companies and an associated drive toward increased innovation in the medium-term

Critics of LCRs argue that the instrument leads to inefficient allocation of resources, a slowdown of the build out of renewable energy capacity, and higher cost, and they question the medium-term gains. It is observed that most countries base their policy decisions on political grounds, rather than on economic and empirical analyses. According to footnote 92, for example, the claimed potential positive spillover effects in the medium term have not been modeled or demonstrated in any country so far. Furthermore, many LCRs used in national RE policies are often poorly designed for national value creation and fail to score well against the identified basic conditions. In many countries, LCR rates seem very high, which increases their trade-distorting impact and the inefficient allocation of resources. In some countries, LCRs focus on components that have low learning-by-doing potential, or on non-infant industries, which does not necessarily lead to innovation or reduced cost. An observation is that many LCRs focus on upstream activities such as manufacturing and neglect the great potential in the value of downstream services, such as engineering, installation, and maintenance.

On the legal side, it is important to consider that support schemes with LCRs for renewable energy are generally prohibited under the World Trade Organization (WTO law, which seems to

---

[101]ICTSD (2013). *Local Content Requirements and the Renewable Energy Industry: A Good Match?*–www.ictsd.org.

be a problem of feed-in tariff schemes. In 2009, the Canadian province of Ontario passed the Green Energy and Green Economy Act, which introduced a feed-in tariff program, coupled with an LCR. Under the LCR, firms are required to use a certain percentage of locally manufactured material for wind and solar projects to receive government support. Ontario did not phase in the LCR gradually and as a result, retail electricity prices increased by more than 17% in 2010. The EU joined Japan in its WTO complaint against Ontario's scheme. The WTO Appellate Body concluded that these requirements were indeed a violation of the legal provisions in the GATT[102] and TRIMs[103] agreement.

Procurement tenders that contain LCRs, however, will hardly be disciplined by the WTO and may therefore be permissible.

7.1.2 Discussion

Given the demographics and sizeable investments in new and renewable energy capacity in many countries, it is understandable that governments look for ways to get the maximum economic benefit, particularly the creation of jobs, off the investments in renewable energy. However, there is very little empirical evidence that the pathways generally chosen, with more focus on upstream than downstream elements of the value chain, are effective in the intended way. Especially in recent years, the competition in the manufacturing of solar photovoltaic cells and modules has dramatically increased. This competition has led to the current affordability and competitiveness of solar energy, but the margins in manufacturing are currently razor-thin. To survive in such an environment, the leading companies have a massive scale and use cutting-edge technology. Their scale, with production capacities of several gigawatts per annum, also gives them a competitive advantage in the management of their supply chain. In a largely commoditized market environment, manufacturers can only survive if their cost structure matches or surpasses that of the competition. Since scale is a very important parameter, for any new factory, a clear pathway to achieving sufficient scale to be competitive on a global level in the medium term is

---

[102]General Agreement on Tariffs and Trade.
[103]Trade Related Investment Measures.

required. Governments should therefore resist the temptation of stimulating new manufacturing capacity in their countries through LCRs, without policy measures that create a market of sufficient size. Furthermore, even though the idea from the outset is to limit the LCRs in time until the home-grown industry can compete with international players, subsidies tend to become very sticky and politically sensitive. Local players become addicted to their protected environment and often form powerful lobbies when the government decides to stop or reduce the subsidy or protected environment. Another consideration should be that there are more jobs in the manufacture of support structures, inverters, cables, junction boxes, and batteries and the engineering, installation, and management of systems. Therefore, governments are better advised to focus on those areas if local manufacturing is desired.

7.2 Nascent Manufacturing Sector

According to the World Bank, the share of manufacturing in sub-Saharan Africa's GDP has been declining from 15% in 1981 to 10% in 2015. A large part of that decline can be attributed to the increased competition from Asia, which has become the so-called "workshop for the world" for many goods. However, Africa's downward trend has been reversed, and since 2009 an upward trend can be observed, which is linked to Africa's tremendous economic growth in recent years.

Several African countries have started solar module factories. There are broadly three potential advantages of local manufacturing: job creation, the opportunity to export and earn foreign exchange, and cost savings. According to research by the author (Frank Wouters), by mid-2017, 10 African countries were home to 21 module factories with a cumulative manufacturing capacity of around 900 MW per annum. It should be noted that it is very hard to obtain accurate information about these companies and their actual production, but the real capacity number should not be very far off, although the actually produced number of modules could be much less. In 2016, more than 60,000 MW were produced globally, so the African continent produced less than 1.5% of that. One also has to realize that 50% of the global volume, or 30,000 MW, was produced by seven Chinese manufacturers, which form the so-called Silicon Module Super League (SMSL). The initial industry group members were Canadian Solar, Hanwha Q Cells, JA Solar, Jinko Solar, and Trina Solar. Yingli Green and GCL, the world's largest solar polycrystalline silicon manufacturers, joined the SMSL in mid-2016. Each of these companies produces on average more than five times the combined production of the entire African continent. Without the innovation capacity of these giants and their market might in the supply chain, it is hard to compete on cost or quality. Nine of the 21 African manufacturers have production capacities smaller than 10 MW/a, which classifies them as niche producers or R&D facilities. Also, half of the African production capacity was in the hands of Chinese manufacturers, for example, incentivized to produce for the local market in South Africa. It is doubtful that a small African factory with a me-too product can be competitive on international markets or be cheaper than a low-cost Chinese imported product. Also, solar modules are typically long-lived; so there is not a lot of business in repairs or maintenance.

But there is more than the solar module that goes into a system. The overall solar PV system is formed by connecting the module to balance-of-systems (BoS) components. These include the inverter, charge controller, cables, connectors, junction boxes, batteries, trackers, and control systems. And in dusty environments, one may need tools to clean the modules, and certain pay-as-you-go systems incorporate GSM-operated switches. Also, there is a range of appliances specifically designed for solar applications. These include LED lights, mobile phone chargers, TVs, fans, and refrigerators. There may be a good business case to manufacturer those in Africa.

7.2.1   Fosera

Germany is known for its innovative Mittelstand, consisting of many family-owned SMEs. According to Prof. Hermann Simon, 48% of the mid-sized world market leaders come from Germany.[104] These firms, which he calls "Hidden Champions," are part of what makes German economic growth more robust than other countries. Several elements characterize such companies, among which are stable leadership and focus on certain products and services that they do best. And it is the focus that encourages these companies to look for opportunities abroad, because the German market alone may not be big enough. A prime example of such a company is Fosera, which was founded by Prof. Dr.-Ing. Peter Adelmann, a global champion of the solar off-grid sector. Dr. Adelmann, a good friend of the author (Frank Wouters), whom he has known since the early 1990s, has been successfully doing business in rural areas of developing countries for decades. He is a serial entrepreneur who has founded dozens of companies. Fosera manufactures a range of solar-enabled products that provide electricity services to the people who are not connected to the grid, such as solar systems for lighting, phone charging, and various other uses, depending on the system size. Adelmann handed the reins of

[104]https://hbr.org/2017/05/why-germany-still-has-so-many-middle-class-manufacturing-jobs.

the company to his daughter a few years ago, but it is not in his nature to let it go completely; so he supports her and the company where he can. He has a clear view on the rationale for local manufacturing in Africa. First of all, local presence enables a better market entry. Second, most products require repair or maintenance and if that can be done in the country, it offers a clear advantage in terms of cost and time. The third element is that local manufacturing requires lower working capital. Last, local manufacturing enables "closing the loop" by recycling the products or elements thereof. The latter is particularly relevant for batteries, which can almost entirely be recycled, which has a local cost advantage.

One of the largest markets in Africa is Ethiopia. Ethiopia is the second most populous country in Africa with 86 million inhabitants and large parts of the country are not covered by the grid. Only Nigeria has more people. Market size is an obvious criterion when deciding on a location for manufacturing, and when Fosera was looking for opportunities to expand in Africa, Ethiopia was a natural choice. According to Adelmann, selling solar equipment in Ethiopia is easy, but many other things pose significant challenges. To start with, there is a currency challenge in Ethiopia. Revenue is mostly in Ethiopian Birr, but many items and materials for the product line need to be imported; hence, foreign exchange is required. Unfortunately, exchanging local currency into foreign exchange is heavily regulated by the central bank and can take more than 12 months. After an approval has been secured, the money has to be spent within three days. This bizarre situation forces the management to have on average open applications with up to 10 local banks for such transactions. The next hurdle is getting raw materials and goods through customs, which is not straightforward. If countries want to have the benefits of local manufacturing, in particular the creation of a local industry, such challenges need to be addressed.

One complicating issue, prevalent in many African countries, is that solar equipment and systems are often exempted from import duty. However, raw materials are usually not, making it difficult for domestic manufacturers of solar equipment to compete with cheaper imports. Although it is laudable that governments promote the use of solar equipment by exempting them from import duties, they effectively put up barriers for local manufacturers. A smart solution for this conundrum is required, e.g., by waiving import duties on raw materials that are being used to manufacture solar equipment, or providing other tax benefits to compensate for this disadvantage.

Fosera also considered setting up manufacturing in Mozambique. However, due to Mozambique's communist past, the entrepreneurial spirit in the country is work in progress, and bureaucracy is omnipresent. According to Adelmann, the country is also spoilt by donors, who have provided generous hand-outs in the past and continue to do so. As a result, the overall environment has proven not to be conducive enough for manufacturing; so the company decided to try its luck elsewhere.

Since 2014, Fosera has a manufacturing cooperation with Solinc in Kenya, Ethiopia's neighbor. Solinc was originally set up by the Dutch module manufacturer Ubbink in 2009, together with ABM Ltd from Kenya. Since 2015, Solinc is majority Kenyan owned. It manufactures a range of solar modules for the Kenyan market but is also exporting to neighboring Tanzania and Uganda. One of Solinc's shareholders is Chloride Exide, a local battery manufacturer, with a vested interest in off-grid energy systems. The recycling rate of lead acid batteries in Kenya is more than 90%, which is advantageous both for the environment and for the cost structure of the manufacturer based in the country. Apart from PV modules, Solinc manufactures a range of products designed and engineered by Fosera, including their range of solar home systems that can be extended when demand increases, by connecting to systems with a parallel cable. The manufacturing capacity of Solinc is a tiny fraction of the gigawatt size of the big seven in China. However, Solinc makes a European designed quality product for the local market and offers a range of products beyond the module only.

Fosera's example shows that innovative companies with a clear strategy can produce solar products in Africa and make a profit. With growing markets and supporting regulatory environments that are favorable for local manufacturing, it should be possible for African producers to thrive. The urge to only focus on the manufacturing of solar modules should be avoided; there may be more margins in other products and systems. Also, the repair and replacement market is larger for shorter-lived products than solar modules, which are very long-lived.

7.2.2   Solar Manufacturing in the Middle East

Several years ago, the expectations of an imminent solar market breakthrough in the Middle East were high. These expectations were among other things spurred by ambitious announcements in Abu Dhabi and Riyadh, by the likes of Masdar and the King Abdullah Centre for Atomic and Renewable Energy (K.A.CARE). Especially, Saudi Arabia's initial plan of more than 41 GW solar by 2032 caused a lot of excitement, and many companies started looking into the solar supply chain. Large investments were made in the production of polysilicon, wafers, cells, and modules in the region. In 2011, Saudi Arabia's oil behemoth Aramco announced a joint venture with Shell's Showa to manufacture thin-film solar modules in the Kingdom of Saudi Arabia, and Qatar Solar Technologies (QSTec), part of the Qatar Foundation, became the largest single equity shareholder in Germany's Solarworld. QSTec has constructed an 8000 metric ton per annum polysilicon factory in Ras Laffan, 45 minutes from Qatar's capital Doha, with the option to expand into wafer and cell manufacturing. And from 2009 to 2012, the author (Frank Wouters) was responsible for Masdar PV, which was part of Abu Dhabi's portfolio of renewable energy investments. Masdar PV operated a thin-film solar PV factory in Germany with the ambition to set up local manufacturing in the United Arab Emirates. However, the regional market took longer to come to scale, and in the meantime, the tremendous growth in the global PV industry reduced manufacturing margins substantially. Also, the previously expensive silicon technology has reduced in price such that it is harder for thin-film technologies to compete, since the efficiency of crystalline silicon cells is higher. As a result, Masdar's plans to produce solar modules in Abu Dhabi were shelved in 2011.

However, recent developments have changed the market outlook for both projects and manufacturing in the region considerably. First of all, Dubai and Abu Dhabi are now home to the lowest-cost solar energy projects in the world. Dubai's 800 MW and Abu Dhabi's 1200 MW Solar PV projects cost less than 3 cents per kWh and are cheaper than any other kind of generation by some margin. As a result, there is a growing pipeline of projects and the Middle East solar market seems to finally kick off in a substantial way. Moreover, both Europe and the United States charge import duties on China-made modules and cells, which is not the case for the products made in the Middle East. So, in recent times, several initiatives have emerged to look into the manufacturing of solar products again, several of them in Dubai.

7.2.3   Noor Solar Technologies

The Bahmani Group, headed by Gholam and Behnam Bahmani, headquartered in Dubai, has been importing, exporting and assembling generators, air conditioners, air coolers, heaters, and spare parts thereof since 1986. After they recognized the demand for solar energy products in the region, the Bahmanis decided to embark on an ambitious manufacturing program and established Noor Solar Technologies. They hired two industry veterans, Jafar Javadi and Luan Kuqi, to head the company. In the beginning of the project, Javadi and Kuqi, CEO and COO, respectively, were still based in Germany and Hong Kong and commuted to Dubai. However, with the facility shaping up toward the end of 2016, both moved permanently to Dubai. Together with the shareholders, they devised a strategy to manufacture a range of solar products. The core is a solar module factory with an initial capacity of 200 MW per year. In addition, they established cooperation with LTI from Germany to manufacture solar inverters. They will also manufacture the support structures for solar projects at their plant in Dubai.

The module manufacturing line has been supplied by the Italian supplier Ecoprogetti, which has provided solar manufacturing equipment since 1998. Ecoprogetti first assembled the production line at their premises in Italy, which enabled thorough testing, but also certification. The line was consequently disassembled, sent to Dubai, and assembled again and is now fully operational for both modules and inverters. But the company will not stop there. When the author (Frank Wouters) visited the impressive factory in July 2017, Luan showed him the large area next to the factory, which is earmarked for expansion. Depending on demand, Noor Solar Technology will reach manufacturing capacity of between 500 and 1000 MW per annum, which will enable them to better manage the supply chain and reduce the cost.

Dubai's strategic location, excellent infrastructure (sea and airport), and conducive business climate are all favorable arguments for manufacturing. If they can offer a product with differentiating qualities, for example, a "desert module," specially designed to withstand the harsh environment in the Middle East and North Africa, it should be possible to compete with Chinese manufacturers, who offer generic global commodity products.

8

**Current and Future Solar Programs
in Africa and in the Middle East**

8.1 Introduction

The incredibly rapid cost decline of solar PV in the past years has caused many governments to rethink their energy policies. When the author (Frank Wouters) was in charge of Abu Dhabi's solar energy projects from 2009 to 2012, solar energy was the most expensive form of electricity. Now, just a few years later, it is by some margin the cheapest form of electricity. And this is the case in many other countries across Africa and the Middle East, most of which have an excellent solar resource. The challenge is to turn this new reality into actionable policy, because solar energy is a lot more versatile and can be deployed in many more ways than, for example, coal or large-scale hydropower. Governments now need to implement appropriate policies that not only involve subsidies, which were necessary in the past, but create a conducive environment for businesses to play their role, because the large portion of Africa's population that is not served by the grid—and will not be served in the foreseeable future—can be served by the private sector in more innovative ways than governments typically do. Pay-as-you-go solar companies, companies selling solar lanterns and other products, and operators of solar-based mini-grids are a case in point. A regulatory environment that creates incentives and removes barriers for such initiatives is arguably more effective than subsidized programs, which have a checkered past of successes and failures, especially in African rural areas. For on-grid power, the government can still play a central role, e.g., in competitive procurement of large solar farms. Well-structured approaches such as the IFC's Scaling Solar program has led to incredibly affordable solar electricity in Zambia, despite the country's low credit rating and associated risk profile.

This section describes governments' approaches to advance solar agendas, now and in the future.

The Sun Is Rising in Africa and the Middle East: On the Road to a Solar Energy Future
Peter F. Varadi, Frank Wouters, and Allan R. Hoffman
Copyright © 2018 Pan Stanford Publishing Pte. Ltd.
ISBN 978-981-4774-89-5 (Paperback), 978-1-351-00732-0 (eBook)
www.panstanford.com

8.2 Africa

After nightfall, most of the African continent is still in the dark, more than a century after the light bulb was invented as it was shown at the beginning of this book (Chapter 1.2). According to the World Bank, some 25 countries in sub-Saharan Africa are facing an electricity crisis evidenced by rolling blackouts. On average, the blackouts cost more than 2% of GDP, not factoring in the potential for additional economic activity if the supply of generation capacity would match real demand. But worse still, 76% of sub-Saharan Africans, 590 million people in total, are not affected by the blackouts, since they don't have access to electricity at all.

The comparison with South Asia, which has similar per capita incomes, is particularly striking. In 1970, Sub-Saharan Africa had almost three times the electricity-generating capacity per capita as South Asia. In 2000, South Asia had left sub-Saharan Africa far behind—with almost twice the generation capacity per capita.[105]

However, Africa, which in the 1980s and 1990s used to be labeled the "lost continent" by many, has been cast in global spotlight thanks to its impressive economic growth since 2000, leading to what people now call the "African century." Most of the world's fastest growing economies since then have been in Africa and many countries are finally on the way to become middle-income countries, with extreme poverty declining and the middle class growing. However, the necessary infrastructure for sustained economic growth is inadequate and its development is lagging. Also, the total number of people without access to modern energy is rising because electrification efforts are outpaced by population growth.

For an informed discussion, one must differentiate between sub-Saharan Africa and the countries in the Middle East and North Africa, the MENA countries that are bordering the Mediterranean Sea, and the Arabian Gulf. In the MENA region, more than 99% of the population has access to electricity, whereby more than half of sub-Saharan Africans lack access.

---

[105]Cecilia M. Briceno-Garmendia, Vivien Foster, eds. (2010). *Africa's Infrastructure: A Time for Transformation*, AFD/WB, Washington, D.C.

8.2.1 Electricity in Sub-Saharan Africa

Cameroon, Côte d'Ivoire, Gabon, Ghana, Namibia, Senegal, and South Africa have electricity access rates exceeding 50%, and the rest has an average electrification rate of 20%. According to McKinsey,[106] it takes 25 years to progress from a 20% electrification rate to 80% electrification rate. Furthermore, there is a strong link between electrification and GDP per capita, typically with steep growth once a country reaches access rates above 80%. It is therefore now time for decisive action. Although the lack of power holds back growth, African resourcefulness always finds solutions, albeit at huge cost. Many people and businesses have their own generators and it is estimated that in Kenya more than half of businesses own generators, while the estimated number of generators in Nigeria amounts to a staggering 60 million. The cost of electricity from such diesel and petrol generators is on average four times the cost of grid electricity. In Nigeria, diesel fuel is a leading expense for the major African mobile phone companies, representing up to 60% of operators' network costs.[107] African enterprises have identified unreliable power supply as the most pressing obstacle to the growth of their businesses, ahead of access to finance, red tape, or corruption.[108]

The installed capacity in sub-Saharan Africa in 2012 amounted to 90 GW, half of which is in South Africa. Forty-five percent of this capacity is coal (mainly South Africa), 22% hydro, 17% oil (both more evenly spread), and 14% gas (mainly Nigeria). Affordability is a critical issue, since electricity prices are typically very high by world standards, despite often being subsidized. Electricity consumption per capita is, on average, less than 1,000 kWh per year, which is less than that needed to power two 50 W light bulbs continuously.

Table 8.1 shows the installed solar PV capacity of African countries at the end of 2016, amounting to 2,491.3 MW, as

---

[106]Antonio Castellano, et al. (2015). *Brighter Africa: The Growth Potential of the Sub-Saharan Electricity Sector*. McKinsey and Company.

[107]Emmanuel Okwuke (February 2014). Nigerian telcos spend N10b yearly on diesel to power base stations—Airtel boss, dailyindependentnig.com.

[108]Fatih Birol, et al., IEA (2014). *Africa Energy Outlook: A Focus on Energy Prospects in Sub-Saharan Africa.*

reported by IRENA. The capacity includes grid-connected as well as off-grid applications and is based on overall data on import of solar modules, which is a pretty accurate measure of what has entered the countries. Although it is more difficult to extract exact data on off-grid applications than the grid-connected capacity, IRENA estimates that off-grid solar PV amounts to 648.5 MW, or roughly a quarter of the overall solar PV generation capacity. It is noteworthy that Algeria, Egypt, Ethiopia, Kenya, Morocco, South Africa, and Uganda all had more than 20 MW of off-grid capacity installed at the end of 2016. Some of that capacity is found in sizeable mini-grids (Egypt, Morocco, Algeria), while in South Africa off-grid may not be truly off-grid (standalone) either, but backup systems and panels installed to reduce electricity bills.

Table 8.1 Installed solar PV capacity in African countries at end of 2016

Country	MW	Country	MW
South Africa	1,544.0	Sudan	8.8
Algeria	224.6	Niger	8.0
Reunion	181.0	Mali	6.0
Senegal	54.0	Zimbabwe	6.0
Egypt	39.0	Malawi	5.7
Tunisia	37.5	Benin	5.0
Mauritania	34.6	Libya	5.0
Uganda	34.0	Zambia	3.0
Namibia	31.6	Burundi	2.6
Ethiopia	30.0	Togo	2.4
Ghana	27.5	Seychelles	2.0
Kenya	25.5	Guinea	1.8
Mauritius	24.7	Botswana	1.7
Morocco	20.6	Somalia	1.3
Nigeria	18.4	Swaziland	0.9
Angola	15.0	Congo Rep	0.5
Mayotte	13.1	Lesotho	0.4

(*Continued*)

Table 8.1 (*Continued*)

Country	MW	Country	MW
Mozambique	13.0	Guinea Bissau	0.3
Cabo Verde	11.0	Central African Republic	0.3
Madagascar	11.0	Djibouti	0.3
Tanzania	10.6	Liberia	0.2
Burkina Faso	10.0	South Sudan	0.2
Rwanda	9.4	Sierra Leone	0.0
Cameroon	9.0	TOTAL	2,491.3

Source: IRENA.

Table 8.1 shows that at the end of 2016, only 40% of all African countries had more than 10 MW of solar PV capacity and only Réunion, Algeria, and South Africa had more than 100 MW. One must consider that Réunion, a small island east of Madagascar, is one of the overseas departments of France. It is also an outermost region of the European Union and, as an overseas department of France, part of the Eurozone. It, therefore, has the same favorable regulatory framework for Solar PV as France, hence the comparatively large PV capacity.

8.2.2 Nigeria

Nigeria, Africa's most populated country, has had a checkered past regarding its recent energy supply history. Part of OPEC, and one of Africa's major oil and gas producers, one would expect it to have a stable supply of electricity. However, the country's 182 million people had an average grid-connected capacity of 3,100 MW practically available to them in 2015, which is less than one-third of the estimated real demand.[109] In April 2016, the system completely collapsed, and for several days there was no grid-connected capacity at all. To keep businesses running, more than 60 million diesel generators keep the lights on, but at tremendous cost to the economy. One of the solutions the government has recognized is to tap into the vast renewable

---

[109]https://www.pwc.com/gx/en/growth-markets-centre/assets/pdf/powering-nigeria-future.pdf.

energy resources the country has. The Nigerian Federal Ministry of Power's National Policy on Renewable Energy and Energy Efficiency set a target for renewable energy (excluding large hydropower) of 8% of Nigeria's total generation capacity by 2020. This translates to roughly 2,000 MW and is to be obtained through competitive procurement for utility-scale renewable energy projects, and net metering and feed-in tariffs for smaller renewable energy systems. This three-pronged approach makes a lot of sense given the current state of play and market dynamics in the country.

However, let us first look at the country's electricity sector. In 2005, Nigeria laid down the structure of the sector in the Electric Power Sector Reform Act. The Act unbundles the previously vertically integrated sector into its main components of generation, transmission, and distribution and also founded an independent regulator, the Nigerian Electricity Regulatory Commission (NERC). Although the distribution system has been split into 16 geographically separate entities that were consequently privatized, the sector is still in a transitional stage toward being fully liberalized. Until market forces will completely determine supply and demand dynamics and prices, the Nigeria Bulk Electricity Trading Company (NBET) is the sole purchaser of power at the transmission level. By law, NBET shall conduct competitive procurement according to a procedure established by NERC to which extent Regulations for the Procurement of Generation Capacity were issued in 2014. Further relevant regulations issued by NERC include Regulations on Feed-In Tariff for Renewable Energy Sourced Electricity in Nigeria and the Regulations for Mini-Grids (2016). The three-pronged approach for grid-connected renewable energy in Nigeria consists of large systems that are connected to the high-voltage transmission grid, medium-sized systems that are connected to the medium voltage distribution networks, and smaller systems that can be built under a net metering scheme. Furthermore, there are provisions for other systems such as mini-grids, mainly for applications where the grid doesn't reach.

8.2.2.1 Large grid-connected projects in Nigeria

As per NERC's regulations, NBET, the sole purchaser of power in Nigeria, is obliged to conduct competitive procurement for bulk power. The way to do that is to organize an auction, which is rapidly becoming the standard for procuring renewable power capacity all over the world. Well-structured auctions are transparent and lead to best value for money. The author (Frank Wouters) was part of the team that worked with NBET in 2016 to prepare such an auction. Although the Ministry of Power, Works and Housing decided to award power purchase agreements to 14 developers that had previously negotiated directly with the government, and push the first auction a bit further into the future, the "threat" of an auction significantly reduced the tariff.

8.2.2.2 Feed-in tariffs

The Nigerian government understood that a healthy mix of larger systems connected to the transmission grid and medium-sized systems connected to the distribution grid could complement each other. However, the auction mechanism that is used for larger capacities is not suitable for smaller systems due to the high transaction costs. Such systems, for solar PV in the ange of 1 to 5 MW, are best served with a feed-in tariff mechanism based on the successful German design. Nigeria adopted the main elements of that design:

- a 20-year power purchase agreement
- a fixed tariff that enables the developer to recover the cost of the investment
- a quick and uncomplicated licensing procedure
- priority access to the grid

The overall amount of renewable energy capacity that is allocated to be funded using feed-in tariffs amounts to 2,000 MW, but this includes all renewables, not just solar.

8.2.2.3 Net metering

There is a large market for distributing generation, systems that are smaller than 1 MW and that are being installed by consumers to reinforce the unreliable grid electricity. For this market segment, the Regulator has introduced a net metering scheme. In this scheme, any surplus electricity that the owner of the system cannot consume will be fed back to the grid at the same rate the consumer pays when buying from the grid. In terms of complexity of the contractual and technical arrangements, this is the simplest of the three.

8.2.2.4 Other solar applications

There are a multitude of solar applications that can serve the hitherto unserved population in the rural areas. These include solar home systems, solar water pumps, and mini-grids. The government realizes the limitations of its own institutions and is building a regulatory framework that supports the private sector in serving the rural population. One such regulation, the NERC Regulation for mini-grids, lays down the opportunities and obligations of developers of mini-grids, including tariff setting. The operator of a mini-grid can either use the so-called Multi Year Tariff Order (MYTO), which applies to all grid-connected generation, or agree with the majority of the customers on a different tariff, if that is required to make a project feasible.

8.2.2.5 Discussion

The Nigerian government has made some significant steps in formulating a regulatory environment for renewable energy. The three-pronged approach for procuring grid-connected capacity is smart and recognizes the most appropriate approaches for different system sizes. The Nigerian economy badly needs additional power and competitive renewable energy is a natural candidate, especially given the excellent solar resource in the country. However, mid 2017 not many megawatts have been installed for a variety of reasons. It took considerable time to put the regulatory framework in place and the lack clarity in the interim period created confusion among potential investors. Second, there are serious deficits in tariff collection at the level of the distribution companies. They consequently fail to fully reimburse NBET for the power supplied to them, which makes the entire system underfunded. Until this problem is solved, it will be difficult for the government to embark on an ambitious solar expansion program, despite its solid design and good intentions.

8.2.3 Uganda

Uganda is a landlocked and, for Africa, relatively small (241,038 km^2) country. It is surrounded by five countries (Rwanda, Tanzania, Sudan, Kenya, and the Democratic Republic of the Congo). The population is 37,873,253. With 215 kWh/a, it has the lowest per capita electricity consumption in the world. Uganda's electric energy generation capacity as of December 2016 was 958 MW. Table 8.2 shows the electricity generation mix.

Table 8.2 Uganda's electricity generation mix

Electricity generation	Fuel	Number of systems	Capacity MW	Total MW
Hydroelectric	Run of river	13	0.3–250	695.4
Thermal	Oil (diesel)	3	1.5–80	131.5
Thermal	Bagasse	5	1.6–52	119.6
Thermal	Solar and oil (diesel)	1	1.6	1.6
Solar	Solar (PV) energy	1	10	10
TOTAL:				958.1

Hydropower is the major source of electricity in Uganda. The country may have some oil production starting in 2020 but as of now, no oil or natural gas is being produced. Other than Tanzania, the four surrounding countries have also no oil or natural gas production. Tanzania has natural gas production but it is not exporting any of that. Without indigenous coal reserves, Uganda have to ship coal and diesel for electric power generation from a port through Kenya and the generated electricity is expensive.

Uganda has several rivers that can be used to generate more electricity. The other, limited, source of renewable energy available in Uganda is bagasse, a byproduct of sugar cane production. The development of additional hydropower systems could double or triple the present capacity and requires a sizable amount of money. The government of Uganda (GoU) and its Electricity Regulatory Agency (ERA) decided to raise capital based on the German Feed-in Tariff system, which was already used very successfully in many other countries. ERA established a Renewable Energy Feed-in Tariff (REFiT) system in 2007, which was initially only for new hydropower and bagasse cogeneration. This program was not successful, mainly because a tariff alone in a country such as Uganda is not enough. The risk that an investor does not get paid for his generated electricity in Uganda is much higher than similar investments in countries such as France or even in South Africa. Uganda was not alone in this conundrum. Most of the sub-Saharan countries had and still have this problem.

To solve the problem of how to best attract investments into those Sub-Saharan countries, the Advisory Group on Energy and Climate Change (AGECC) of the Secretary General of the United Nations had asked Deutsche Bank's experts to present new concepts for promoting renewable energy investments in developing regions. As a result, the "Global Energy Transfer Feed-in Tariff" (GET FiT) scheme was developed by experts of Deutsche Bank Climate Change Advisors and presented in January 2010. Deutsche Bank, in cooperation with the German development bank KfW, developed a concept for a GET FiT pilot program and decided that it should be tried in Uganda. The GET FiT program in Uganda was launched by KfW in 2013. Chapter 4.6 of this book describes in detail the program and its results.

As described in Chapter 4.6, the GET FiT program in Uganda planned 13 hydropower projects and 1 bagasse project. The bagasse power plant was completed in mid-2015. However, at the end of 2016, none of the 13 hydropower projects had been completed. Two of a total of 12.6 MW were completed during the first quarter of 2017, seven of them are still in the construction phase, and four have not even been planned yet. The problem with hydropower plants is that it takes a long time from planning until they become operational. Also, it is difficult to accurately predict the time for developing such projects because of difficulties in construction that are hard to accurately foresee from the outset. As a result, Uganda has the potential to establish more small hydropower plants, but they cannot solve immediate electricity shortage problems.

With solar PV prices drastically reduced, the price of the electricity produced by PV systems became very competitive with hydropower systems. PV electricity is obviously much cheaper than that produced by diesel generators. Furthermore, the big advantage of PV is that large systems can be installed and become operational in 6 months and have a guaranteed performance for at least 20 years. Realizing this, ERA asked KfW to include PV systems in their GET FiT program. The first 10 MW PV systems were installed and became operational in November 2016. The second 10 MW systems are expected to be operational during Q3 of 2017. At present, three more PV systems totaling 80 MW are planned and four totaling 500 MW are in early planning stage.

8.2.4 Namibia

Namibia's Atlantic Ocean coastline is about 1,572 km long and the country's land area is 823,290 km^2. It has 2.5 million inhabitants (2017), so it ranks among the least densely populated countries in the world. Table 8.3 indicates how much electric energy is produced by NamPower, the national power utility company of Namibia, by its various generating systems.

Table 8.3 Electricity production in Namibia

Electricity generation	Fuel	Number of systems	Capacity (MW)	Remarks
Hydropower "Ruacana"	Run of river	1	347	
Thermal "van Eck"	Coal	1	120	Because of the expensive coal, only a 30 MW section is operating
Oil "Short-term peak demand"	Diesel	4 × 6.5	26	Standby station for the coastal area
Various PV systems	Sunshine	>10	50	Estimated as per mid-2017
Total:			**543**	
In operation:			**427**	

NamPower'selectricity production is mainly reliant on one hydropower plant that is not sufficient to provide enough electricity for the country. Namibia is therefore extremely dependent on electric energy imports. About 60% of its need is imported, mostly from South Africa. This explains the high electricity charges, especially for business customers, and is also the reason for frequent blackouts and power interruptions. The main electricity grid is from the country's northern border with Angola, where the country's only hydroelectric power plant is located, 1,600 km to its southern border with South Africa, from where the majority of its electricity is supplied. The problem is that there is only a small coal-powered electricity plant in the capital, Windhoek, and a small auxiliary station in the only Atlantic port near the southern border. Therefore, almost the entire country has to be supplied using power feeders branching off from this main backbone. Another problem of Namibia is that a large number of the population is scattered in the large country and cannot be reached even by such feeders.

Namibia has to solve two problems as soon as possible. One problem is to provide inexpensive and stable electricity for customers able to utilize the existing grid. The other problem is to provide electricity for those who cannot be reached by the grid.

Obviously, Namibia has to reduce its dependence on imported electricity, but first the situation with the traditional sources of electricity has to be reviewed.

The Ruacana hydropower plant is the only one to supply electricity for the country. In 2016, NamPower increased its power generation capacity by 15 MW from 332 MW to 347 MW, which is the plant's limit. At present, only one additional hydropower plant is planned, the Baynes Hydropower Project, which will have an installed capacity of 600 MW. It will be located along the Cunene River, which originates in Angola and is 200 km downstream of Ruacana. This plan is now only in the feasibility phase and when that is completed, it will take 6 years to be in operation. Obviously, the Baynes hydroelectric power plant will not help to solve Namibia's power problem in the short to medium term.

Namibia has no indigenous oil, gas, or coal. Imports have proven to be expensive, so Namibia has to find alternatives to develop domestic power sources to decrease its expensive electric power imports and its dependence on energy imports. Although

Namibia is the World's fourth largest uranium producer, it is unlikely that it will be able to obtain and operate a nuclear electric power station in the foreseeable future.

8.2.4.1 Utilization of renewable energy to produce electricity

To tap into the great renewable energy resources in the country, the so-called REFiT program was initiated by the Electricity Control Board (ECB) and was operationalized in October 2015. The ECB's plan is to add 170 MW renewable electricity capacity to the national grid. In the Namibia REFiT program, the NamPower prepared standard Power Purchase Agreements (PPAs) for biomass, concentrated solar power, solar PV and wind energy systems (hydropower was not included).[110] The ECB recently also introduced a net metering system for PV systems. Net metering allows residential and commercial customers who generate their own electricity from solar power to feed electricity they do not use back into the grid. If the home is net-metered, the electricity meter will run backward to provide a credit against what electricity is consumed at night or other periods where the home's electricity use exceeds the system's output.

8.2.4.2 Biomass

An area of the combined size of Germany, Austria, and Switzerland has been taken over by bushes and trees, called "bush encroachment," caused by overgrazing and climate change. To make use of this resource, NamPower conducted a pre-feasibility study on the so-called bush-to-electricity concept. The excess woody biomass is estimated to be 200 million tons per annum. The pre-feasibility study indicated that a 20 MW biomass power plant would require an annual supply of 150,000 tons of chipped biomass that can be harvested from many regions within a 35 km radius. It was found that sufficient bush biomass is available to supply 10 plants of 20 MW each.

It was described in Chapter 4.6 that NamPower requested KfW to undertake a detailed design and implementation readiness study to develop a program concept for a GET FiT program

---

[110]http://www.nampower.com.na/refit.

"bush-to-electricity" in Namibia. The study shall be financed by the Government of Germany and it will be carried out in the course of 2017.

NamPower is also contemplating putting up so-called hybrid power installations, where electricity is generated from solar during the day and from wood during the night.

8.2.4.3   Wind

On Namibia's Atlantic coast, the wind resource is excellent for wind power. A private consortium led by Quantum Power recently signed a PPA with NamPower for a 44 MW wind farm, near Namibia's costal town Lüderitz. The first 5 MW wind power system is reported to be operational during the second half of 2017.

8.2.4.4   Concentrated Solar Power (CSP)

CSP systems operate only with direct sunshine. Over a large part of the country, Namibia is mostly cloudless and the solar irradiation level is the second highest in the world at 3,000 kWh/m^2/a. Therefore, it would be an ideal location for CSP systems. Today the cost of CSP electricity is somewhat more expensive than produced by stationary and even single-axis tracking PV systems, but the advantage of CSP systems is that they can provide electricity around the clock with integrated thermal storage.

Despite the fact that the neighboring South Africa already has seven CSP systems with a total capacity of 600 MW in operation (seven more are planned) and that NamPower's ReFiT program already has a PPA for CSP systems, as of today no concrete plan exists for the construction of a CSP system.

8.2.4.5   PV Systems

After KfW successfully developed the GET FiT program in Uganda, they offered Namibia to introduce the same program, starting with PV systems. As described in Chapter 4.6, Namibia informed KfW that its REFiT and the net metering program was very successful for PV systems and therefore they did not need further assistance in that area.

Among Namibia's REFiT programs, the solar PV component became very successful. The reason is that a large population is living in the areas not served by the grid, and in the areas where electricity is available, its supply is not reliable because the country's electricity-generating capacity covers only slightly more than one-half of its annual electricity requirements. Namibia is therefore highly dependent on imports.

What is the advantage of PV? First, PV can be used for decentralized electricity generation anywhere and in any size, from a few watts to multi-megawatt capacity. Furthermore, PV became cheap a few years ago and can be deployed and put into operation very fast. Second, people in Namibia knew from experience that PV is a reliable source of electricity. The utilization of PV in Namibia started as far back as the 1980s. Small PV systems were used mostly by farmers for water pumping and for lighting and radios. Therefore, people in Namibia have 35 years of experience with the reliability of PV systems.

The first company to import PV modules and install them was Solar Age Namibia (PTY) Ltd.[111] The company was started in 1989 by Conrad Roedern. Now, the Renewable Energy Industry Association of Namibia[112] has 30 members.

Figure 8.1 Karibib 5 MW single-axis solar PV power plant.[113]

As part of the REFiT program, an interim program of 70 MW was established, focusing on small-scale RE producers. The

---

[111]http://www.solarage.com/.
[112]http://www.reiaon.com/.
[113]Courtesy of Bruno Reihl, CEO, HopSol AG (www.hopsol.com).

PV REFiT program limits the production capacity to 5 MW per project therefore 14 PV projects were commissioned.

Table 8.4 shows a list of the solar PV units installed, operational and connected to the grid under the REFiT program.

Table 8.4 REFiT PV power plants in Namibia

Power plant	Community	Type	Capacity	Year completed
HopSol Otjiwarongo	Otjiwarongo	Fixed, thin-film CIGS	5 MW	2015
InnoSun Omburu	Omaruru	Thin-film CIGS, single-axis tracking	4.5 MW	2015
InnoSun Osona	Okahandja	Thin-film CIGS, single-axis tracking	4.5 MW	2016
Ejuva I	Gobabis	Single-axis tracking	5 MW	2016
Ejuva II	Gobabis	Single-axis tracking	5 MW	2016
HopSol Otjozondjupa	Grootfontein	Thin-film CdTe, single-axis tracking	5 MW	2016
AEE Power Ventures Rosh Pinah Power Plant	Rosh Pinah	Single-axis tracking	5 MW	2017
HopSol Karibib	Karibib	Single-axis tracking	5 MW	2017

Figure 8.2 Otjozondjupa Solar Park, 5 MW single-axis thin-film (CdTe) solar PV power plant.[114]

---

[114]Courtesy of Bruno Reihl, CEO, HopSol AG (www.hopsol.com).

8.2.4.6 Commercial and other organizations

With low prices and 25-year guarantee for PV systems, their reliability compared with the electric power system, and installation times of only months, many commercial and other organizations established their own large PV systems. One of them was Namibian Breweries, with their own plant producing 34% of their electricity demand with a roof-mounted 1.1 MW PV-generating system (Fig. 8.3).

Figure 8.3 Namibia's largest roof-mounted grid-connected solar PV installation of 1.1 MWp located at Namibia Breweries in Windhoek, installed by Solar Age Namibia (Pty) Ltd.[115]

There are more examples:

- A Namibian supermarket chain equipped 13 of their branches with a total of 1.5 MW.
- PV systems have been installed on the headquarters of NamPower and the Ministry of Environment and Tourism.
- Several mines have PV systems, e.g., the Kalahari mine for manganese and rare metals has an 800 kW PV system.
- The 292 kW off-grid PV system at Gam has a battery storage to make its electricity available 24 hours (see Fig. 8.4).

Even small users find PV useful for protecting themselves from high electricity costs and power failures. An example is a shop in Walvis Bay (Fig. 8.5).

---

[115]Courtesy of Solar Age Namibia (PTY) Ltd (www.solarage.com).

Figure 8.4 Off-grid 292 kW solar PV plant, 576 individual 2 V batteries with a combined capacity of approximately 2.5 MWh, and its control room at Gam.[116]

Figure 8.5 Shop, Walvis Bay.[117]

---

[116]Courtesy of Bruno Reihl, CEO, HopSol AG (www.hopsol.com).
[117]Courtesy of Bruno Reihl, CEO, HopSol AG (www.hopsol.com).

As already mentioned, even in the 1980s, when PV prices were high, many farmers started to use solar PV to pump water and provide for lighting and radio. Since the prices of PV systems came down, the utilization of PV became widespread. However, since there are no records on these systems, it is difficult to estimate the overall PV capacity in this market segment, but it is expected to be substantial.

8.2.4.7 Summary

It is an "educated guess" that as of September 2017 at least 50 MW PV capacity was installed and in operation in Namibia. By the end of 2018, the combined, grid and off-grid PV capacity will be close to 100 MW. This is a very high number considering that the country only has 2.5 million inhabitants.

8.2.5 Senegal

Senegal is a West-African country whose power production is almost entirely based on thermal power plants, which mostly run on oil and natural gas. According to the African Development Bank, about 61% of the population has access to electricity. While this is a decent share for African comparison, most of these electrified people live in urban and more densely populated areas—as indicated by an urban electrification rate of nearly 75%. Yet, about half of Senegal's population lives in rural areas where only 18% have access to the electricity grid.

Senegal's electricity sector has been liberalized and to a large extent privatized for almost 20 years. The responsible body in Senegal's electricity sector, ASER, has championed a concessionary model for rural electrification called Programme prioritaire d'électrification rurale (PPER), in which companies, after a competitive process, were given the mandate to provide electricity to 10 different regions by either conventional means such as grid extension or the installation of mini-grids. In areas outside the concessions, local initiatives can get electrified by applying for support through an approach called Électrification Rurale d'Initiative Locale (Rural Electrification upon Local Initiative). These areas are typically more remote; so decentralized solutions such as mini-grids or solar home systems are likely to be the least-cost approach. Between 2005 and 2009, many villages were served through such initiatives; so there is about 10-year experience in certain areas with this approach. Although the assumption has always been that rural electrification is important for the development of people and communities, not much scientific data are available to support this, especially since, for instance, solar home systems typically provide basic electricity for consumptive use and rarely for productive use. Nonetheless, according to the research carried out by Ruhr University Bochum, Germany,[118] personal discussions with people in non-electrified rural areas reveal that electrification is one of their most urgent needs—most importantly because of electric lighting. Rural dwellers in Africa often say that the much brighter electric lighting increases both the perceived and the objective security. It appears straightforward that this also changes the perception of life and social as well as economic decision-making. The German researchers tried to quantify less tangible effects of electrification going beyond the narrow focus on the Millennium Development Goals.

8.2.5.1 Impact of solar home systems in Senegal

German researchers[116] carried out a survey among households in 56 villages in the remote Casamance area in southern Senegal that were part of the first phase of a solar home system (SHS) dissemination program. For the dissemination of these SHS, the program pursued a fee-for-service approach. The disseminated SHS consist of a 55 Wp solar panel, a battery, a charge controller, and four compact fluorescent lamps. Some people had small TVs. The payments consist of an initial down-payment and monthly subscription payments. In the villages where the systems were offered, approximately half of all the households had rented a system after 2 years. An adjacent area with villages where the second phase of the program in a similar fashion would be offered served as a comparison to the electrified villages. A scientifically sound methodology was used to create like-for-like comparisons in terms of ethnic make-up, wealth, and education

[118]Gunther Bensch, Jörg Peters, Maximiliane Sievert (2012). *Fear of the Dark? How Access to Electric Lighting Affects Security Attitudes and Nighttime Activities in Rural Senegal*, ISBN 978-3-86788-424-2.

in the households studied. Although other consumptive uses such as radio and TV were observed, it was mainly lighting that makes up the attractiveness of electricity. The aim of the research was to examine how lighting changes attitudes and behavior of people, taking into account that qualitative interview partners highlighted security and comfort issues related to lighting. In terms of security, sometimes people mentioned the risk of being robbed, but others were a little vaguer and mentioned the "sense of living in obscurity." The researchers wanted to measure how daily life of people changes with the advent of electric light. One important indicator is the time kids spend at home studying. On security, the researchers asked respondents whether they were comfortable if their kids were outside after nightfall and whether they themselves would go outside after nightfall.

The results on studying were significant. In electrified households, children study 25% longer than in non-electrified households of comparable nature, and most of that extra time is spent after nightfall. On security, the results were not as striking. Although the people in electrified households perceive a higher sense of security, there was little empirical data to show that they are attacked less by robbers or animals than in non-electrified cases. However, they would leave their kids play outside slightly longer because of perceived higher security.

8.2.5.2 Solar energy in the Middle East and North Africa

In the Middle East and North Africa (MENA) region, more than 99% of the population has access to electricity, which is a different situation compared with sub-Saharan countries. Also, several MENA countries have substantial oil and gas reserves; so the drivers to deploy renewable energy projects are different. Some countries, such as Morocco, import most of their fossil fuels, whereas other countries that have domestic fossil fuels would like to diversify their energy mix to reduce growing domestic use of own oil and gas. Although for a number of years, announcements about renewable energy projects and policies have been made, so far there has been little real activity on the ground. However, a number of developments point to the fact that finally the turning point has been reached, and the MENA region will start accelerating the introduction of renewable energy. There are areas with a very good wind resource, such as

Morocco's Atlantic coast, or Egypt's Zafarana Region, but other areas have only moderate wind speeds. However, solar irradiation is excellent throughout the region. The recently increasing cost-competitiveness makes for a very compelling case to deploy solar PV at a big scale. In many countries, the demand for electricity keeps growing at rates of between 6% and 8% and the demand profile, due to air-conditioning, coincides perfectly with the supply of solar energy.

According to the Middle East Solar Industry Association MESIA,[119] the MENA PV pipeline is more than 4,000 MW in 2017 alone. The main countries where that will happen are Morocco, Algeria, the UAE, Tunisia, Jordan, Oman, Egypt, Saudi Arabia, and Kuwait.

8.2.6 Morocco

Morocco is the only North African country with no indigenous oil resources and is the largest energy importer in the region, with 96% of its energy needs being sourced externally. The leading supplier of Morocco's energy requirements is Saudi Arabia at 48%. By 2009, Morocco's energy bill had reached $7.3 billion and electricity demand is projected to quadruple by 2030. The government set a goal of reaching 42% of installed capacity (or 6,000 MW) from renewable energy from hydro, wind, and solar by 2020, while doubling overall capacity.[120] Solar power in Morocco is enabled by the high rate of solar insolation, about 3,000 hours per year of sunshine but up to 3,600 hours in the desert. Through the Morocco Solar Plan (MSP), the government aims to install 2,000 MW of solar capacity by 2020, contributing around 14% of the energy mix in the country's electricity supply. The plan calls for the construction of five solar complexes, requiring an estimated investment of $9 billion. This program, called NOOR, will be implemented by the construction of solar power plants in Ouarzazate (510 MW CSP and 70 MW PV), Tafilalt and Atlas (300 MW PV), Midelt (300 MW CSP and 300 MW PV), Laayoune and Boujdour (100 MW PV), Tata (300 MW CSP and 300 MW in PV), and solar power plants in the economic

[119]http://www.mesia.com/wp-content/uploads/MESIA-OUTLOOK-2017-Lowres. compressed.pdf.
[120]http://www.nortonrosefulbright.com/knowledge/publications/66419/ renewable-energy-in-morocco.

zones (150 MW PV). The procurement program is managed by the Moroccan Agency for Solar Energy (MASEN), which is a public-private agency dedicated to implementing the Moroccan Solar Plan and the promotion of solar energy by developing solar power projects, contributing to the development of national expertise and proposing regional and national plans on solar energy.

The Moroccan procurement program has achieved excellent results by awarding power purchase agreements for CSP projects in Ouarzazate, which achieved world-record low prices for concentrated solar power at the time. Recently, MASEN has launched a dedicated PV program called NOOR PV I, and the first phase entails three projects with a combined capacity of 170 MW. The pre-qualification took place in December 2015 and the projects are expected to go online in 2017.

Why Morocco's solar ambitions have been more successful than other countries can be attributed to MASEN, which was set up next to the national utility ONEE and empowered by direct support of King Mohammed VI of Morocco. MASEN has been the main interface with the lenders, such as the World Bank and the European Investment Bank, and has achieved a very favorable financing package, bringing the cost down substantially. In other countries, where the incumbent utility has been a central player in the renewable energy strategy, uptake of renewable energy projects has been considerably slower.

An interesting approach by the Moroccan government through MASEN is the development of hybrid PV-CSP (HSP) projects. The first phase of Noor-Midelt is such an approach. The project consists of two separate plants, each with 150–190 MW CSP capacity and a minimum of 5 hours of thermal storage. The capacity of the PV component, which is expected to provide daytime generation, is left to the bidders' discretion, but cannot exceed night-time net capacity from CSP by more than 20%. The advantages of this approach are twofold. First, a combined PV-CSP hybrid system with integrated solar thermal storage is capable of providing dispatchable power not only during the day but also during the evening and even throughout the night. Second, the low-cost PV in the mix reduces the average price per kWh compared with a CSP-only project.

8.2.7 Egypt

Egypt's fantastic history is marked not only by the pharaohs and their pyramids and temples but also by one of the earliest industrial solar power plants that was built in the country. More than 100 years ago, in 1912 in Meadi, parabolic solar collectors were established in a small farming community by Frank Shuman, a Philadelphia inventor, solar visionary, and business entrepreneur. The parabolic troughs were used for producing steam, which drove large water pumps, pumping 6,000 gallons of water per minute to vast areas of the arid desert land.

In more recent times, Egypt has been struggling to meet its growing power demand with domestic oil and gas. Since Egypt has great solar and wind resources, it was a natural choice for the government to embark on an ambitious renewable energy journey and it plans to raise its share of renewable energy capacity to 20% by 2022. In 2014, Egypt announced a renewable Feed-in Tariff (FIT) program, with a target of 2.3 GW of solar PV by 2017, of which 2 GW will be centralized PV power plants and 300 MW will be distributed PV installations under 500 kW. The first solar tender was strongly oversubscribed. After initial delays, among other things due to a floating of the currency late 2016, and a reduced feed-in tariff (FiT) (from 14.34 to 8.40 U.S. cents per kWh), the program is now moving forward with a smaller number of developers. The land at the project sites in Benban/Aswan and Zafarana has been allocated to the awarded bidders. Close to 1,500 MW of projects are expected to reach financial close in 2017.

The program is supported by multilateral financing institutions such as the IFC, EBRD, and OPIC.

Apart from these large grid-connected projects, Egypt also has a sizable share of the population that is currently not connected to the grid. For them, solar energy is an attractive solution and the government has recently been working on electrifying these villages, primarily in the south of the country, either through programs distributing solar home systems or solar-assisted mini-grids.

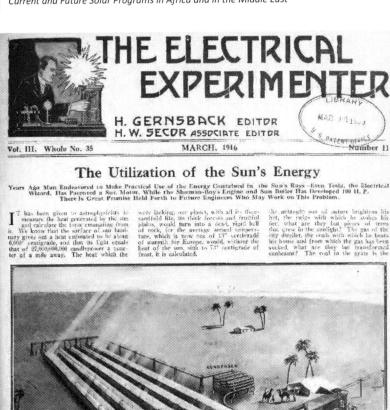

Figure 8.6 The Shuman-Boys solar collector array at Meadi, Egypt, 2013.[121]

---

[121]https://en.wikipedia.org/w/index.php?title=File:The_Electrical_Experimenter,_Volume_3.pdf&page=643.

8.3   The Middle East

8.3.1   Jordan

Jordan has a fantastic solar and wind resource and has embarked upon an ambitious renewable energy program a few years ago, mainly to become less dependent on unreliable and expensive imported fossil fuels. It has been looking at increasing its share of renewables to 10% by 2020, which includes a goal of reaching 600 MW of solar PV. In May 2015, the Ministry of Energy and Mineral Resources (MEMR) awarded power purchase agreement to developers of four 50 MW solar PV projects, each around 6 U.S. cents per kWh, which is lower than for conventional generation. In comparison, a recently awarded oil shale project in Jordan required a tariff of around 13.5 U.S. cents per kWh, more than double the tariff for solar energy. Although the oil shale is an indigenous resource, the CO_2 emissions per kWh, a major factor in the decision making, are the highest of any technology.

Jordan not only procures large grid-connected solar PV capacity but also allows companies to directly procure bulk solar power from generators elsewhere. The power is delivered to them via the NEPCO grid through a procedure called power wheeling. The main driver for this is commercial; companies pay a lot for power and can save money buying cheaper solar energy. In 2016 approximately 80 MW of power wheeling projects came online. Companies are also allowed to install PV panels on their premises in a net metering scheme, whereby they would be reimbursed for the surplus electricity fed into the distribution grid at the same tariff they pay. In 2016, approximately 50 MW of net metering projects were realized.

8.3.2   United Arab Emirates

In 2006, Abu Dhabi, the largest of the UAE's seven emirates, launched the ambitious Masdar initiative. Masdar, which means "source" in Arabic, aims to build an ecosystem around sustainable energy, enabling Abu Dhabi to remain a central player in the global energy economy. Abu Dhabi has large oil and gas reserves but realizes that over time fossil fuels will be replaced with sustainable energy. Rather than fighting it, or not being part of its development, Abu Dhabi has made the strategic choice to pro-actively invest in sustainable energy technologies, real estate, and projects.

The author (Frank Wouters) joined Masdar in 2009 and was responsible, as director of Masdar's Clean Energy Unit, for a portfolio of clean energy investments all over the world, worth more than $8 billion. These included a factory producing PV modules in Germany, factories producing wind turbines in India and Finland, several large concentrated solar power plants in Spain and the UAE, and London Array, the largest offshore wind farm in the world.

Masdar's flagship project is Masdar City, a carbon-neutral part of Abu Dhabi close to the international airport. Masdar City is powered by a 10 MW grid-connected ground-based PV system, supported by a 1 MW rooftop system. The 10 MW system is a combination of First Solar and Suntech modules, and the rooftop was supplied by SunEdison. It is ironic that these three companies were all once the largest solar PV companies in the world. However, two of them have had to file for Chapter 11 protection since, which shows that despite the steady and spectacular growth of solar PV in the past decade, the market development has been rather turbulent and not an easy ride for manufacturers.

After the 10 MW system in Masdar City was built, Masdar started developing Shams I, a concentrated solar power plant with a capacity of 100 MW, to be located in Madinet Zayed, some 140 km from Abu Dhabi. As president of the company, the author (Frank Wouters) led the construction, which involved 15,000,000 man-hours. The plant was based on the conventional parabolic trough technology and was supplied by Abengoa. Masdar's co-investors were Abengoa from Spain and Total, the French oil company, which is now the majority owner of Sunpower. The plant was completed in 2013 and has been outperforming its planned specifications ever since. Although the original plan was to build three more CSP plants, the massive price drop of solar PV has changed the immediate priority for Abu Dhabi.

In parallel to Abu Dhabi's efforts, neighboring Dubai, which doesn't have significant oil or gas reserves, started looking at solar as well. The Maktoum Solar Park on the outskirts of Dubai started with a 13 MW solar PV installation in 2014. Initially, Dubai's solar ambitions were relatively modest, but just before COP21 in November 2015, the Dubai Clean Energy Strategy 2050 was announced, which greatly increased the emirate's clean energy targets. Dubai now aims to cover 7% of its power with clean energy by 2020, 25% by 2030, and 75% by 2050. The previous target was 15% by 2030. The main reason for the increased ambition level was the fact that DEWA, the utility company in Dubai, achieved a world record of 5.84 cents per kWh for their 200 MW solar PV project in the second phase of the Maktoum Solar Park. Since Dubai has to import most of its energy, solar energy turned out to be the cheapest form of power in Dubai. DEWA decided to also introduce a rooftop program, based on net metering. Since, contrary to most other places in the Gulf (including Abu Dhabi), electricity is not subsidized in Dubai, commercial and industrial users of electricity can save money using this scheme. By mid-2016, an estimated 350 MW of rooftop solar PV projects using this scheme are under way, and according to DEWA, the grid can easily absorb 2,500 MW of distributed PV.

However, DEWA was in the global headlines again early 2016, when the results of the third phase of the Maktoum Solar Park, a huge project of 800 MW, were announced. The winning bid, offered by the author's (Frank Wouters) former colleagues at Masdar, came in at a world-record low price of 2.99 cents per kWh. Astonishingly, this tariff is unsubsidized, using commercial debt. Solar energy is not only the cheapest form of electricity in Dubai; it is now so by a wide margin! Of course, it helps that the off-taker has a solid credit rating and that loans are still very cheap. However, the same argument is valid for other energy investments.

The original size of the Maktoum Solar Park was increased from 1 GW to 3 GW in early 2015. At the end of 2015, the planned size increased to 5 GW, reflecting the increased appetite for solar PV.

Abu Dhabi, in the meantime, did not sleep but issued a call for tender for a 350 MW solar PV project in 2016, to be built in Sweihan. The tender was left open for developers to offer a larger project as long as it would fit on the site and would be more cost-effective. The result was another world-record low price of

2.42 cent per kWh this time. One should bear in mind that this tariff is for 8 months, while in the four hot summer months a 60% bonus is awarded, which means that it is on average roughly the same as the Dubai tariff of slightly below 3 cents per kWh. However, the winning consortium of Marubeni from Japan and Chinese module supplier Jinko Solar managed to squeeze 1.2 GW onto this site. In Dubai, the size of the project, the credit rating of the off-taker, the availability of cheap credit, and the quality of the bidding consortium all contribute to this low tariff.

8.3.3 Saudi Arabia

Saudi Arabia has initiated a number of very ambitious renewable energy plans in the recent past. The kingdom produces much of its electricity by burning oil, a practice that most countries abandoned long ago. Most of Saudi Arabia's power plants are inefficient, as are its air conditioners, which consumed 70 % of the nation's electricity in 2013. Although the kingdom has just 30 million people, it is the world's sixth-largest consumer of oil. The Saudis burn approximately 25% of the oil they produce, and their domestic consumption has been rising at an alarming 7% a year, nearly three times the rate of population growth. According to a widely read report by Chatham House,[122] a British think tank, from December 2011, domestic consumption could eat into Saudi oil exports by 2021 if this trend continues. At this rate of demand growth, national consumption will have doubled in a decade. On a "business as usual" projection, this would jeopardize the country's ability to export to global markets and render the kingdom a net oil *importer* by 2038. Given its dependence on oil export revenues, the inability to expand exports would have a dramatic effect on the economy and the government's ability to spend on domestic welfare and services.

This plain fact has been the main driver behind the government's plans to diversify the energy mix, and the late King Abdullah bin Abdulaziz Al Saud, who passed away in 2015, founded the King Abdullah Centre for Atomic and Renewable

---

[122]https://www.chathamhouse.org/sites/files/chathamhouse/public/Research/Energy,%20Environment%20and%20Development/1211pr_lahn_stevens.pdf.

Energy (KA CARE) by Royal Decree in 2010. KA CARE, based in Riyadh, published a scenario in which, by 2032, Saudi Arabia would generate 50% of all electricity from non-fossil fuels. KA CARE's scenario incorporated nuclear, solar, wind, waste-to-energy, and geothermal. Next to hydrocarbons, nuclear, and other renewable sources, solar generation capacity would expand to 41 GW, of which 16 GW solar PV and 25 GW by concentrated solar power. According to their predictions, solar PV would meet total daytime demand year-round; concentrated solar power, with storage, would meet the maximum demand difference between photovoltaic and base load technologies; and hydrocarbons would meet the remaining demand. This scenario was widely published and created a lot of investors' attention for this promising market, which would be the largest solar PV market in the region. However, 7 years after KA CARE's inception, many things have happened, but we are yet to see the promising large-scale initiatives that would really start the market.

However, in April 2016, Saudi Arabia launched Vision 2030. As part of that vision, the Kingdom plans to install 9.5 GW of renewable energy, about a quarter of the previous target. The goals reflect the ambition to overhaul the economy of Saudi Arabia, including privatizing the electricity sector and divesting a stake in the state-owned oil company Aramco to diversify away from fossil fuels as a primary revenue source.

With the recently achieved low prices for solar PV, it is just a matter of time before Saudi Arabia starts deploying solar at a large scale. It makes economic sense and it creates employment opportunities, especially if the kingdom decides to start manufacturing, which is viable because of the market scale. This is an important aspect, since approximately 50% of Saudi's young adults are currently unemployed.

When the market finally kicks off, one of the biggest firms waiting in the wings is ACWA Power International, which is based in Riyadh and Dubai and owns and operates power and desalination plants in the Middle East, Africa, and Southeast Asia. In the past few years, ACWA Power has signed contracts to produce solar power in several countries—places where the price of conventional electricity is higher than in Saudi Arabia. ACWA won the bid to build a solar farm in Dubai, the 200 MW phase II of the Maktoum Solar Park, which set a world record at 5.84 cents per kWh in 2015, and the firm won the contracts for the ambitious Noor solar complex in Ouarzazate and Midelt in Morocco, a combination of CSP and PV.

Although ACWA Power did not win the bid for the next phase in Dubai, the 800 MW phase III, they came close and are among the most aggressive investors in large-scale solar power projects in the world. As soon as projects materialize in Saudi, experienced developers such as ACWA and others are ready to engage and transform the fossil fuel giant into a green energy powerhouse.

Early 2017, Saudi Arabia finally started the long-awaited renewable energy program. A call for tender was launched for a 300 MW solar PV project and a 400 MW wind energy project, and the results of the competitive procurement process will be announced before the end of 2017. Indications are that a tariff of less than 2 cents per kWh is going to be awarded, which implies another world record for solar energy. Given the size of the market and the promise of regular annual auctions for years to come, many companies are showing interest and the shortlist of qualified bidders was not short at all. The Kingdom has implemented a few crucial structural improvements, which are meant to streamline the renewable energy program. First, the responsibility for water and electricity has now been put within the energy ministry, removing any ambiguity about who is in charge. Second, there is a lot more clarity about the roles and responsibilities of the various government agencies.

8.4   Into the Future

The past decade has shown unprecedented growth in Africa and the Middle East, with more people than ever lifted out of extreme poverty. However, the number of people who live in rural areas, where the grid does not reach, has still grown and conventional energy solutions struggle to provide a compelling case for the future. The great solar resource, the increasing cost-competitiveness of solar solutions, and innovative business models shine a bright light on Africa's energy future. Large-scale solar farms will feed the grids of African and Middle-Eastern countries, buildings will have solar roofs and energy-poor people in the rural areas will have access to affordable and healthy solar technologies that can help them out of poverty. The once-neglected solar markets can turn into engines for green economic growth.

Epilogue

An energy transition that took its first tentative steps in the latter part of the 20th century is now unfolding rapidly in the 21st century. It will have a major impact on Africa and the Middle East along with every other part of the world. It is a transition from dependence on carbon-based fuels such as coal, oil, and natural gas to the utilization of renewable energy technologies such as solar, wind, biomass, geothermal, hydropower, and ocean technologies. All, but geothermal, which is derived from the radioactive decay heat in the core of the earth, and tidal energy caused by the moon, are direct or indirect forms of solar energy. Just as we have experienced a fossil fuel era for the past few hundred years—today the world is still more than 80% dependent on such fuels—we are now embarking on a solar energy era that taps into the enormous amounts of energy received by the earth from its sun 150 million kilometers away. To put this in context, while the earth intercepts approximately 6 million exajoules of solar radiation each year (1 exajoule = 10^{18} joules), and the total global energy consumption is about 600 exajoules, the fraction of the sun's radiated energy intercepted by the earth's disk is only 4 parts in 10 billion. The issue before us is how to utilize this diffuse energy source cost-effectively and meet, in an environmentally friendly way, the needs of an expanding global population.

We are transitioning from relying on ever-scarcer sources of fossil energy to an era of unlimited, clean, and cheap energy, brought about by modern technology. This transition, which can also be seen as an energy revolution, has major implications for bringing energy services not only to urban and peri-urban areas of Africa and the Middle East but also to those rural, off-grid areas currently without access to electricity. Both Africa and the Middle East are blessed with enormous solar resources, which are just beginning to be tapped, providing an opportunity to improve the lives of hundreds of millions of people. Efficient and cost-effective solar solutions and novel business models enable previously unserved people to leapfrog straight into the future of energy. This book explores some of these opportunities that will transform Africa and the Middle East in the decades ahead. It is an exciting time in the energy history of the world, and Africa and the Middle East will be important playing fields in creating that new history.

Glossary

ACP	African, Caribbean, and Pacific Group of States
ADB	African Development Bank
AGECC	Advisory Group on Energy and Climate Change
ASTAE	Asia Sustainable and Alternative Energy Program (World Bank)
CIGS	Copper indium gallium diselenide
COP 17	Conference of the Parties 17
CPV	Concentrating photovoltaics
CSP	Concentrating solar power
DRC	Democratic Republic of Congo
DOE	U.S. Department of Energy
EAPP	Eastern Africa Power Pool
ECOWAS	Economic Community of West African States
ECREEE	ECOWAS Centre for Renewable Energy and Energy Efficiency
ERA	Electricity Regulatory Authority
ERERA	ECOWAS Regional Electricity Regulatory Authority
ESA	European Space Agency
ESMAP	World Bank's Energy Sector Management Assistance Program
EU	European Union
FED	European Development Fund
FiT	Feed in Tariff
FMG	Financierings Maatschappij voor Ontwikkelingslanden N.V. (Netherlands Development Finance Company)
GCC	Gulf Cooperation Council
GCCIA	Gulf Cooperation Council Interconnection Authority

GEF	Global Environment Facility
GET FiT	Global Energy Transfer Feed in Tariff
GFITPPM	GET FiT "Premium Payment"
GW	Gigawatt
HCPV	High Concentration PV
HSP	Hybrid Solar Power System
ICTSD	International Centre for Trade and Sustainable Development
IEC	International Electrotechnical Commission
IECEE	IEC System of Conformity Assessment Schemes for Electrotechnical Equipment and Components
IFC	International Finance Corporation
ILO	International Labour Organization
IPP	Independent Power Producer
IRENA	International Renewble Energy Agency
KfW	Kreditanstalt Fuer Wiederaufbau (KfW Development Bank)
LCOE	Levelized cost of electricity
LCPV	Low concentration PV
LCR	Local Content Requirements
LED	Light emitting diode
mbgl	Meters below ground level
MENA	Middle East and North Africa
MIGA	Multilateral Investment Guarantee Agency
MW	Megawatt
NASA	National Aeronautics and Space Administration
NEPCO	National Electric Power Company
NDC	Nationally determined contributions
NGO	Non-governmental organization
OPEC	Organization of Petroleum Exporting Countries
OPIC	U.S. Overseas Private Investment Corp.

PAYGO	pay-as-you-go
PIDA	Program for Infrastructure Development in Africa
PPP	Public Private Partnership
PRG	Partial Risk Guarantee
PV	Photovoltaic
PV-GAP	Photovoltaic Global Approval Program
PVSEC	PV Solar Energy Conference
REFIT	Renewable Energy Feed in Tariff
SHS	Solar Home System
SSA	Sub-Saharan Africa
T&D	Transmission and distribution
TSO	Transmission System Operator
UAE	United Arab Emirates
UNBS	Uganda National Bureau of Standards
UNDP	United Nations Development Program
UNFCCC	United Nations Framework Convention on Climate Change
UREA	Uganda Renewable Energy Association
USAID	US Agency for International Development
WACEC	West Africa Clean Energy Corridor
WAPP	West African Power Pool
WB	World Bank
WTO	World Trade Organization

About the Authors

 **Peter F. Varadi**, after a scientific career, was appointed head of Communication Satellite Corporation's chemistry laboratory in the USA in 1968. In this capacity, he also participated in research on PV solar cells, which were used to power satellites. In 1973, he co-founded Solarex Corporation, USA, to develop the utilization of solar cells for terrestrial applications. Solarex was one of the two companies that pioneered this field. By 1983, it became the largest PV company in the world, when it was sold to AMOCO. Dr. Varadi continued consulting for Solarex for 10 years and then for the European Commission, World Bank, National Renewable Energy Laboratory (NREL), and other organizations. In recognition of his lifelong service to the global PV sector, he received in 2004 the European Photovoltaic Industry Association's John Bonda prize. His book *Sun above the Horizon*, which describes the meteoric rise of the solar industry, was published in 2014. The sequel to the book, *Sun towards High Noon*, was published in 2017.

 **Frank Wouters** has been leading renewable energy projects, transactions, and technology development for over 28 years. He has played a lead role in the development of renewable generation projects valued at over $5 billion. These range from small-scale PV solar electrification in Uganda to the 100 MW Shams I Concentrating Solar Power (CSP) plant in the UAE, and strategic equity investment in the London Array, the world's largest offshore wind project. As deputy director-general of the International Renewable Energy Agency (IRENA), the first global intergovernmental organization dedicated to all renewables, he managed a US$ 350 million IRENA/Abu Dhabi Fund for a development project facility for renewable energy. He appraised over

80 projects a year and recommended projects for funding, including solar PV projects in Africa. Mr. Wouters has served on the boards of several energy companies, including Torresol, where he developed three CSP plants, including the Gemasolar central tower project with molten salt storage. He currently serves as director of the EU GCC Clean Energy Technology Network, a platform that aims to foster clean energy partnerships between Europe and the Gulf, is advising the World Bank on solar energy around the world, serves as senior director—New Energy at Advisian, and is a non-executive board director of Gore Street Capital, London.

 **Allan R. Hoffman** retired in 2012 as senior analyst in the Office of Energy Efficiency and Renewable Energy, U.S. Department of Energy (DOE). He holds a bachelor's degree in engineering physics from Cornell University and a PhD in physics from Brown University. He came to Washington, D.C., in 1974 as a Congressional Fellow of the American Physical Society and subsequently served as a staff scientist for the U.S. Senate Committee on Commerce, Science, and Transportation; director of the Advanced Energy Systems Policy Division of the DOE; assistant director for Industrial Programs, Energy Productivity Center, Mellon Institute; and consultant and senior analyst for the Office of Technology Assessment, U.S. Congress. In 1982, Dr. Hoffman joined the staff of the U.S. National Academy of Sciences/National Research Council, where he served as executive director of the Committee on Science, Engineering and Public Policy. In 1990, he returned to the DOE, where he served as associate and acting deputy assistant secretary for Utility Technologies, with responsibility for a $300 million RD&D program (solar, wind, biomass, geothermal, hydropower, ocean, energy storage, hydrogen, superconductivity). He has also served as the U.S. representative to and vice chairman of the International Energy Agency's Working Party on Renewable Energy. In the decade before his retirement, he pioneered in exploring the linkage between water and energy issues (the water-energy nexus) and helped develop the DOE's offshore wind energy program. Dr. Hoffman is a fellow of the American Physical Society and the American Association for the Advancement of Science.

About the Contributors

Series Editor

Wolfgang Palz has been continuously involved in the development of global photovoltaics (PV) as a scientist and manager for more than 50 years. He is currently the promoter of auto-consumption for PV for providing worldwide access of cheap electricity to all. He has been the leader of PV development in France since the 1970s, when the country was the European leader in the field. As an official of the European Commission in Brussels, he was the manager of PV development in Europe for 20 years. He inspired and supported German initiatives to kick off PV markets, which led to a global explosion of PV investments since 2004. Besides his key role in the development of PV in Europe, Dr. Palz is much connected to the United States, where he worked with NASA and is currently a member of the leadership council of ACORE in Washington, DC. He is equally active in China, where he recently organized the Green Wall Forum on Renewable Energies.

Contributors

Anil Cabraal is a renewable energy consultant working in African and Asian countries. Previously, during his 15 years at the World Bank, he was responsible for rural and renewable energy investments and policy and strategy development, including guiding the World Bank Group renewable energy and energy efficiency scale-up strategy from 2005 to 2010. His work included renewable energy projects and programs in Bangladesh, China, India, Indonesia, the Philippines, the Maldives, Myanmar, Sri Lanka, Kenya, Liberia, Rwanda, Tanzania, and Zambia. While at the World Bank, he, along with the International Finance Corporation, formed the Lighting Africa Program in 2006 and co-led it until 2010. Dr. Cabraal's contributions to photovoltaics include Best Practices for Photovoltaics Household Electrification Programs, Designing Sustainable Off-grid Electrification Programs, and Photovoltaics for Community Service Facilities. He received the Professor Robert Hill award for his contributions to photovoltaics for development at the European Photovoltaics and Solar Energy Conference and Exhibition in 2005. He serves on the Boards of KMRI Lanka (pvt) Ltd., RenewGen (pvt) Ltd., and the Sri Lanka Energy Forum. He earned his doctorate in Agricultural Engineering from the University of Maryland, USA, and BSc in mechanical engineering from the University of Ceylon, Peradeniya, Sri Lanka.

 **Richenda Van Leeuwen** is a globally recognized energy access expert who, from 2010–2016, founded and led Energy Access at the UN Foundation and its engagement with the UN Sustainable Energy for All Initiative. She launched the "Energy Access Practitioner Network," a leading 2,500-member global enterprise network focused on off-grid and mini-grid energy access. She is currently chair, International Institutions at the Global LPG Partnership, focused on cooking energy access. She also serves on the Technical Advisory Group to the Energy Sector Management Assistance Program (ESMAP) at the World Bank. Richenda's deep public and private sector experience includes commercial private equity and impact investment in emerging markets renewable energy companies, including with Good Energies. She was previously CEO of Trickle Up, an international women's microenterprise non-profit and worked in many countries in sub-Saharan Africa, Asia, and the Americas. Richenda currently serves on the boards of SELCO India and Energy 4 Impact. She is also a U.S. Women's "Clean Energy Ambassador" under the Clean Energy Ministerial's C3E initiative and gained both her BSc and MBA degrees at the University of Durham, UK.

Index

Pan Stanford Series on Renewable Energy – Volume 5

Sun above the Horizon
Meteoric Rise of the Solar Industry
Peter F. Varadi
9789814463805 (Hardcover), 9789814613293 (Paperback), 9789814463812 (eBook)
2014

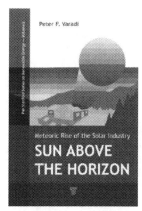

The meteoric rise of the photovoltaic (PV) industry is an incredible story. In 2013, Google's investments in PV systems totaled about half a billion dollars, and Warren Buffet, one of the famous investors, invested US$2.5 billion in the world's largest PV system in California. These gigantic investments by major financial players were made only 40 years after the first two terrestrial PV companies, Solarex and Solar Power Corporation, were formed in the USA. Today, the worldwide capacity of operating PV electric generators equals the capacity of about 25 nuclear power plants. The PV industry is growing at an annual rate of 30%, equivalent to about five new nuclear power plants per year. This book describes how this happened and what lies ahead for PV power generation.

Sun above the Horizon is a must-read, as can be seen from the following citations:

In **The Wall Street Journal**'s August 22–23, 2015, issue, Daniel Yergin writes in his review titled "Power Up": "Solar is growing fantastically," says Dr. Varadi, who chronicles solar's rise in his new book, **Sun above the Horizon**. "Something like this requires time. Shale oil and shale gas had a ready market. When we started, we had no market at all, zero. And the industry had to get to mass production to bring down cost."

Deloitte, the multinational professional services firm, announced their book selection for 2016: "Our featured book is **Sun above the Horizon: Meteoric Rise of the Solar Industry**, 44th book of the Books with Branko program."

"Peter takes you on a fantastic ride through the incredible growth story of modern solar energy from 1973 to the present day."

Frank Wouters
Former Deputy Director-General, International Renewable Energy Agency

"Dr. Varadi's book is a unique contribution to the history of solar PV, an energy technology that is transforming the way we generate and use electricity. I am unaware of any other book that addresses this history as comprehensively as this book does."

Dr. Allan R. Hoffman
Former Associate and Acting Deputy Assistant Secretary for Utility Technologies,
U.S. Department of Energy

"This book provides a unique perspective by one of the pioneers in the PV industry who co-founded the first company that manufactured solar cells for terrestrial applications."

Dr. Denis J. Curtin
Former Chief Operating Officer, XTAR LLC, USA

Pan Stanford Series on Renewable Energy – Volume 7

The U.S. Government and Renewable Energy
A Winding Road
Allan R. Hoffman

9789814745840 (Paperback), 9789814745857 (eBook)
2016

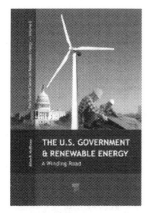

This is a book on how the U.S. and other governments have changed their thinking about energy issues over the past four decades, a change triggered by increasing concern about the role of fossil fuels in global warming and climate change, greater awareness of the risks of nuclear power, and the emergence of viable renewable energy sources. It will enhance understanding of the global energy transition that is finally getting under way in the second decade of the 21st century at an accelerating, even dizzying, pace. Target audiences are the young people who will inherit the transition and shape its future, those in government who currently shape our public policies, and those colleagues, friends, and family members who lived through many of the times and events discussed in the book.

THE U.S. GOVERNMENT
& RENEWABLE ENERGY
A Winding Road

"Hoffman played a substantial role in the development of a wide variety of renewable energy technologies over the past 40 years, while employed at the U.S. Senate, the National Academy of Sciences, and the DOE. Much can be learned by examining the failures as well as the successes. Hoffman tells us what needs to be done for a gentle landing on sustainable technologies with a smart grid. This is an important and necessary path for the nation and the planet."

Emeritus Professor David Hafemeister
California Polytechnic State University, USA, and Author of *Physics of Societal Issues*

"I always had great admiration for those in government who were able to establish programs for the advancement of renewable energy (RE). This is especially true for people in the U.S. government (USG), which was highly influenced by the fossil fuel and nuclear energy industries. Allan R. Hoffman was one of these USG officials who led this effort for many years. He now presents us with this interesting and informative book that describes how RE programs were first formulated and then traveled through a winding road in the USG."

Dr. Peter F. Varadi
Co-founder of Solarex Corporation

"Dr. Allan Hoffman presents a unique personal record of the U.S. energy policy development during four decades. He is one of the top driving forces in this progress and conveys a fascinating description of the successes and disappointments from the inside of the federal government. Earlier than most people, he recognized the potential of renewable energy. He has also been a pioneer in comprehending the water–energy linkage. For anybody who wishes to understand how technology relates to politics, this book is a must-read."

Prof. Gustaf Olsson
Lund University, Sweden

Pan Stanford Series on Renewable Energy – Volume 8

Sun towards High Noon
Solar Power Transforming Our Energy Future

Peter F. Varadi

9789814774178 (Paperback), 9781315196572 (eBook)
2017

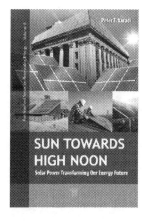

The meteoric expansion of the solar (PV) industry resulted from an incredible reduction in the prices of PV systems—first described in the author's earlier book **Sun above the Horizon**. This book describes how the worldwide PV operational power capacity grew to some 300 GW by the end of 2016. Most of this increased capacity, 250 GW, was installed during the years 2010–2016. Suddenly PV started to affect the traditional generation of electricity. Three practically unlimited new PV markets—residential, commercial, and utility scale—materialized, along with the new PV-oriented financial systems needed to provide the required gargantuan-scale capital. This book also highlights the increasing demand for and the corresponding increased supply of PV cells and modules on four continents and the impact of this PV breakthrough on our lives and future. To present this unparalleled story, the author was helped by the contributions of top experts Wolfgang Palz, Michael Eckhart, Allan Hoffman, Paula Mints, Bill Rever, and John Wohlgemuth.

"This comprehensive and timely book provides the reader with a very thorough technical, regulatory, and financial overview of the global solar (PV) industry. Featuring internationally eminent contributors from the who's who of solar industry experts, this book offers insights, analysis, and background on all the key issues facing this rapidly growing industry. It will be an invaluable reference and resource for scholars, investors, and policymakers dealing with the emerging solar power phenomenon."

Branko Terzic
Atlantic Council and Former Commissioner, U.S. Federal Energy Regulatory Commission

"The long-term welfare of people on our planet depends on an energy system heavily dependent on solar energy. This solar energy handbook presents a well-documented, comprehensive, and insightful view of solar energy's past, present, and future. Its preeminent contributing authors include solar energy pioneers, visionaries, and practitioners who bring a wealth of experience and insights into solar energy markets, financing, policy, and technology."

Karl R. Rábago
Executive Director, Pace Energy and Climate Center,
Elisabeth Haub School of Law,
Pace University, USA

Printed in the United States
by Baker & Taylor Publisher Services